1. My sister Sara, aged 19, Sulejow, 1939.

2. Myself, Moshe Aaron, aged 17, Sulejow, 1939.

3. My brother Szulim, aged 16, Sulejow, 1939.

4. My first attempt at a forest refuge – and an unfriendly boar. *Picture*: A. Pawlen.

5. In my living grave. *Picture*: A. Pawlen.

6. Remains of my caved-in bunker, in Spala forest, near Blogie, 1950.

7. A dry hole covered with leaves and branches: all that remained of my forest well, 1950.

8. Stefan's stable in Chorzew, where I often hid and slept.

9. January 1945, after liberation by the Russians, in my partisan uniform: an army jacket and trousers made from a bag containing supplies dropped from a plane. I am wearing the same boots that saved my leg and, maybe, my life.

10. Back in civilian life, but still wearing the 'supplies' trousers. Taken in photographic studio, Cracow, March 1945.

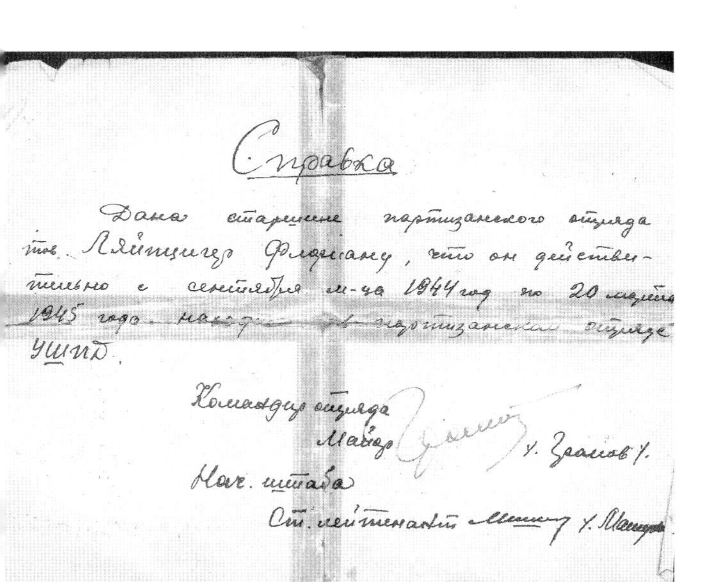

Справка

Дана старшине партизанского отряда тов. Ляйтцигер Фляману, что он действительно с сентября м-ца 1944 год по 20 [illegible] 1945 года нах[illegible] в партизанском отряде УШПД.

Командир отряда
Майор [illegible] Ч. Зрайов Ч.

Нач. штаба
Ст. лейтенант [illegible] Ч. [illegible]

11. Handwritten official partisan declaration confirming my membership of the Russian partisans. Issued in Cracow by the Russian Central Partisan Command, 1945.

12. As a sergeant in the Polish Army, Gorzow, 1947.

13. Addressing a meeting of the regional governing body in Slubice, 1966, in my role as Speaker. The slogan urges Poles to make the nation greater.

14. As president of the Combatants' Association, leading the parade commemorating VE Day, Slubice, May 8 1966.

15. I funded this headstone to replace the wooden crosses that I had built for Janina and Marian's parents' graves in Blogie. I found Janina's name and picture had been added when I visited in 1990.

16. Reunited with my comrade-in-arms, Misha Stepaniuk, the mine-layer. Near Kiev, Ukraine, 1968.

17. My reunion with Stefan and his wife Stasia on my first visit back to Poland, 1990.

slightest noise in the forest, twigs cracking with the movement of deer, the plop of snow falling from the trees. I tried talking to the birds, working out what they were saying to each other and making a similar noise. They responded, but the conversation was above my head.

I fought to survive, but at the back of my mind I was aware that it might not be possible to get through this. There was always the chance I might be discovered by poachers coming through the forest, or I might just die of cold in the night. I knew the temperature in that part of Poland could get to 38 degrees below zero. Outside, my spit would freeze before it hit the ground.

I didn't feel part of the world. A world war was taking place all around me and I knew nothing about what was going on. I couldn't imagine how, even if I survived, I could adjust to the new situation, learning how to deal with people again and being away from nature. I wondered: 'Was I the only Jew left alive?'

Time went by and I noticed that the days were getting longer. When I saw the sun and felt the warmth outside I felt elated. The snow on the trees started melting and dripping down and, with it, my hopes began to grow. The food in my store was getting low and I had to measure careful portions to last until winter was finally over.

It began to rain, quickly dissolving the snow. When I saw the buds I became both excited and frightened. I did not know what lay in store for me, though I was sure the War was still going on. If it had ended, Marian would have come to let me know. I waited several days until the snow in the forest had sufficiently cleared before checking if it was safe to leave. There was still snow on the road, but it was possible to walk.

First I went to check my well. It was full of leaves and branches. I started to clear it out. All the water had to be drained down to the sand base to let clean fresh water come through. It took a while to finish. I then masked the top of the well with large branches. Next day, early in the evening, I set off on my journey to Blogie.

Chicken Soup in the Forest

The forest had become noisy with the chatter and flight of birds making their nests. The air was damp and filled with the smell of rotting leaves. I felt happier, but apprehensive about what lay before me.

At the edge of the forest I realised that it was too early in the day to go to Marian. I looked around, checking for mounds of potatoes in the fields so I knew where to collect them from when I needed to. When it got dark I continued my journey to Marian's house.

Marian was there with Janina. They were surprised and pleased to see me alive, but did not seem happy. I asked how they and the family were and what was happening in the War. The first news was upsetting. Their father had died the previous month. It was now the middle of March 1943.

'And how did you survive in the forest?' they asked. 'Was your bunker warm enough? Did you have enough food?'

'Yes. It was warm. I could close myself in and have enough fresh air through the vent. I was careful about rationing my food and I've still got some left.' I explained how I managed to cope mentally during the winter months and returned the novels and history books they had lent me.

I asked them what day it was. 'Wednesday.' I explained how I had lost track of time. Marian said he would come and visit me at the weekend. It made me feel like I was back in the normal world, thinking about the different days of the week.

Marian gave me a brief account of the progress of the War, how the Germans were losing against the Russians, how Stalingrad had been the turning point. 'I will tell you all the other news when I come and see you.' Their German neighbours had been very good to them when their father died. Willy had been home for Christmas and had become a different boy. He was now a real Nazi. He was still friendly to Marian and Janina.

After supper I returned to my forest, walking for an hour in the dark. In the morning I went to check the water in my well. It was clear and plentiful. I filled my milk can and set about my old routine. It was still very wet on the ground. There was a strong smell of stagnant air. I went to check my second bunker; everything on the outside seemed in order. When I opened the cover I was hit by the dampness. I went in and was glad to find that it was okay to use if needed.

On Saturday I prepared lunch. I boiled a thick barley soup with beans, potatoes, carrots and onions. I calculated correctly, Marian turned up at lunchtime. After enjoying our bowls of soup it was time to talk seriously. I lay down in my sleeping place the opposite way to usual while Marian sat on my seat in front of the fireplace. He filled me in on the War, how the Germans were beaten on two fronts, and were becoming more and more oppressive. They were raiding the villages more frequently, arresting and torturing people. He thought he might have to stay in the forest with me from time to time.

Marian had discovered that Rudi was unhappy with the German occupation. He was taking care of the farm with his mother and sister, though in the coming year he would be old enough and expected to be called up to the army.

Marian asked me if I could make a cross from a birch tree for his father's grave, as I had done for his mother. 'Yes,' I replied. 'I will make it tomorrow and bring it to you in the evening.' Next day, I carried the cross to Blogie and left it in Marian's shed.

I often cooked for Marian on his visits to the forest and he would compliment me on my cuisine. I mentioned that I hoped one day to entertain him and Janina at my own home. Marian said, 'You can entertain us now.'

'What do you mean?'

'If you want to make a meal on Sunday I will bring Janina. She is intrigued with how you manage to live.'

I asked Marian if he could arrange a chicken for me. He said to come in the evening and he would give me one. When

I came to collect it Janina asked if I needed anything else. I said I could do with some eggs, which she gave me. She was excited at the thought of going to my bunker for Sunday lunch.

When growing up, I had often cooked for my grandparents when my mother was away on business and my sister was working. I wanted to cook Janina a traditional Jewish meal. It had to be chicken soup with lockshen.

I already had plenty of saucepans. I put the pan with the chicken on my cooking plate and boiled the soup. I had to make my own lockshen. I mixed up the flour and eggs to make a paste that I rolled out with a large vodka bottle on a chopping board I had pinched from outside a house in the village. I dried the paste and cut it into thin strands of lockshen. I boiled potatoes to go with the meat and prepared some beetroot.

Janina turned up with Marian. This was her first visit. She sat down on a bench I had made. She could not believe what I had done in order to survive. She shook her head and said, 'My God. My God,' and started to cry. Marian had brought a bottle of vodka and we toasted each other.

Joining the Partisans

In our discussions about politics, Marian would say Stalin was as bad as Hitler, which at the time was an unconventional view as most Poles, not just on the left, were sympathetic towards Stalin and were convinced he wanted to save Poland. Marian said Stalin had knifed the Poles in the back by occupying the eastern part of the country when Hitler invaded. This was long before we heard of Russian atrocities: their sending of more than 100,000 Poles, including many Jews, to Siberia, and the systematic murder of thousands of officers, police and intellectuals. Marian wanted to see a return to the centre-right government that ruled before the War.

The best news, Marian explained, was that we were going to organize a partisan group belonging to AK, the Armia Krajowa (Home Army). Its headquarters were in London, where the Polish Government-in-exile co-ordinated strategy. Their aim was to build a new army to liberate Poland and take power after the War.

He wanted to know, 'Would you like to fight the occupiers?'

'Yes.' I had no hesitation. 'With all my strength.'

Winter in the forest had taught me that I did not want to spend another winter like that. I'd been lonely and despairing, missing human company and deprived of news about the world.

Marian told me to come to his place the next day, where he would have a Polish uniform for me. He wanted me to wear it and then go and demand a rifle from a farmer.

It was a very long day. I knew I was entering a new chapter of my life. In my mind I worked out how I was going to force the farmer to give me the rifle.

At sunset I left my camp. At Marian's, Janina fed us. Marian took me to the bedroom and gave me a pale green brass-buttoned Polish Army jacket, which I tried on for size. It was second-hand, hardly worn; a perfect fit, and came with a belt and pistol case, but no pistol. The peaked cap was the right size too. I admired myself in the mirror, feeling like a real officer.

Marian said, 'We'll fill out the case so it will look as if you're carrying a pistol. In the next village, Blogie Rzadowe, I will point out a farm where I know the farmer has a rifle. Demand he gives it to you; tell him you need it to fight the enemy. Ready? Let's go.'

He gave me the name of the farmer and showed me the house. I let myself in. From the hall I heard voices. I knocked on a door and a woman's voice said, 'Come in'. Inside were four middle-aged women tearing goose feathers off their quills. Their conversation ceased when they saw me. A man, also middle-aged, was at the cooker with his back to me.

'I want to speak to Mr Smak. Is that you?'

'Yes, that's me,' said the man, turning round.

'Could we go to another room and talk privately?'

The farmer led the way and asked what he could do for me. He was surprised to see a man in a Polish uniform.

'We know you have a rifle and use it to kill animals in the forest.'

'I have no rifle. Who told you that?'

'We must have it to fight the Germans, that's more important than hunting animals. If the Germans or the police find out you have a rifle you'll be in trouble. I'm asking once more. Give me the rifle.'

'And I'm saying once more I don't have one.'

I put my hand on the pistol case and said loudly, 'Come with me to the forest, I haven't any more time to waste on you.'

'Don't shout, please,' said Smak. 'I will give you the rifle.'

As we went out, I said goodnight to the women. The farmer's wife asked where we were going. I told her he would be back in five minutes.

We went behind his stables. From under the straw he took out a rifle covered with a blanket and gave it to me with a packet of eight bullets. I thanked him and said, 'You are lucky you gave this to me. You would have been in trouble if we'd gone to the forest.' I took the rifle and said goodbye.

Not far from his home Marian was waiting for me. I showed him the rifle and bullets. I wanted to go straight to the forest but Marian said, 'You have to go to my house to leave the uniform behind. I need to return it.'

After I changed, Marian said, 'Tomorrow I'll be at your place and we'll start practising with the rifle. You'll have to learn quickly because when I bring the boys you're going to have to be their teacher.

'I hope you don't mutter your language in your sleep. The boys would kill you if they heard. They'd think you were German.'

I'd been trying hard to forget my Yiddish as Marian had suggested. When I was walking, performing tasks, just thinking or talking to myself, I made sure it was always in Polish. I should be safe.

I had to take precautions with impending guests. I thought it would be wise to hide some personal memorabilia. I cherished my photos of my brother, sister and cousins which had my name on the back of them. I'd also hung on to a school report and my apprenticeship papers.

I found a hand-sized hole high in a pine tree. I wrapped the photos and documents in newspaper and placed them in, covering the space with a patch of bark. It was an easily recognizable tree; I wouldn't forget it. I retrieved everything when I later left the forest, hiding them in the joists of Marian's barn for the remainder of the War.

I left Marian's later that night armed with the rifle. I was very proud and felt more secure. I arrived at my bunker and went inside, putting the rifle on my sleeping place. I made a fire, had a hot drink and prepared to go to sleep.

At lunchtime the next day, Marian arrived and we began practising drill with the rifle. Using all the skills he had learnt as an officer in the army, he made me stand to attention and hold the rifle correctly while on parade. He was teaching me so that I could teach other recruits. He wanted me to be an officer. This carried on for several days. I learned how to take the rifle to pieces and clean it and how to aim. We had bullets but we did not use them; we could not afford to and anyway, it might have attracted unwelcome attention.

At the beginning of April 1943, I was outside my bunker and heard footsteps coming towards me. I grabbed my rifle and prepared to use it. Looking in the direction of the noises I soon made out Marian with a young stranger. Marian laughed when he saw me with my rifle. He was pleased that I was ready to defend myself, and introduced me as Florek, a shortened form of Florian.

'This is Zbyshek Saliniewicz. He ran away from the Baudienst (building service) brigade digging trenches for the Germans on the other side of the River Pilica. He's from Piotrkow and has an uncle in Blogie, though it's not safe for him to stay there. He wants to join the partisans and I want you to teach him how to use that rifle. Let him sleep in your bunker. When more boys arrive he can stay with them in the other bunker.'

Zbyshek was in his early twenties, slim and tall with red hair. He was well spoken and intelligent and had a gentle nature.

Marian left for home and we started to talk about how we would divide up our duties, like bringing water, collecting food, cooking, and military exercises. I told Zbyshek that I came from Sulejow and how I knew Marian. Zbyshek said that his uncle had a farm and a butcher's shop in Blogie. So if we were desperate for some food we could get it, though it was probably better to avoid going there.

I took him to the fields to show him where to collect potatoes and other vegetables. We did daily exercises, preparing for action. The other bunker was almost ready for use; it just needed the wood for the sleeping areas, and blankets. The boys would have to organize that for themselves.

Just over a week later, Marian brought three more boys, all in their twenties: Wolf, his brother Zbik, and Jarek. They also had been digging trenches for the Germans and all three were from Piotrkow. We set about organizing their sleeping places. In the evening, they went to get some blankets in Blogie with Marian, who then had to bring them back in case they got lost.

The boys were excited about their new underground home. It was an adventure. They were happy to be free from working for the Germans.

The First Action

We started exercises straight away the next day with our one rifle. I grew to like the boys and we all got on well. Wolf was short and skinny; he was a tough, blond, streetwise kid who liked to strut around. Zbik was quiet and sensitive, well built, and liked sorting out the domestic arrangements. Jarek was tall, blond and skinny, and very organized.

Wolf suggested that three of us could go to Piotrkow to try and get more weapons. He knew a bar where the Germans went to drink and pick up Polish girls. They often got drunk and we could try and ambush them.

It sounded a good idea, and based on information Wolf and Zbik gave me I devised a plan that Marian accepted. We decided that Wolf, Jarek and myself would set off for Piotrkow the following lunchtime and arrive there that evening.

It was a Friday, market day in Sulejow. We got a lift half way to Piotrkow, walking the rest of the way, another five or six kilometres, and reached Bugay, a suburb of Piotrkow, after sunset. Wolf took us on to a railway track that skirted round the city, allowing us to avoid walking through the streets. From the track we crossed a common to houses on the same street as the bar. We stopped at a large house with gates leading through the middle of the house straight on to the street. We let ourselves quietly inside the gates and kept watch for Germans making their way from the bar. It was fairly early, so Wolf paid a quick visit to a friend nearby to see how things were, making his way behind the houses. It was risky walking the streets after the 10 p.m. curfew.

Around midnight, we could hear a voice singing in German. Looking out, we saw a soldier in uniform; he was swaying as he came down the street towards us. He reached the gate where we were standing. We pretended we were drunk and called out 'Comrade,' in German. He came towards us and gave one of us a hug and said, 'Vodka.'

'Ja, ja,' I answered.

Then Wolf and Jarek took him by the arms, walked him into the gates and behind the house, past the garden and on to the common. I reached for his pistol and took it out of his holster. Then we all ran off. He shouted, 'Help,' but there was no one to hear him.

We went back the same way we came, but had to walk all the way, over 20 kilometres, arriving back in the forest around 5 a.m. Zbik and Zbyshek were asleep. We gave them a fright. They learned they had to be more vigilant.

After lunch, Marian came to the forest and I showed him our booty. He wanted to know everything that happened. We needed more weapons. 'How are we going to get them?' I asked Marian.

'I have a plan,' he said. 'I'll tell you about it soon when I've got it worked out.' He brought us all together. 'We now have a five-man group, and only one rifle and one pistol. Of course, Florek will be your leader and he will carry on taking exercises. We'll have more weapons soon and everyone must know how to use them. There must be discipline in the group and all orders will come from myself to Florek, and then to you. We will see a lot of action and more boys will be joining us. By then you will be ready to take the oath of allegiance to the Armia Krajowa and our country. For the moment, you will have to organize your own food. Florek will show you how. Later, we will be taking food from the Germans and farmers. You will not be staying in the forest all the time, you'll be moving around.'

Before dusk, Marian left for home saying we had to stick to our routine until he returned with orders for our first action. I carried out my duties. The more I got to know the boys, the more I liked them.

It was four days before Marian's return. We were to carry out an action against the police station at Blogie. I was to go to Marian's house and put on the Polish Army uniform before joining the boys at midnight at Anton Zaborowski's farm near Blogie. Then Wolf, armed with the rifle, would follow me into the police station. The rest would wait in the corridor for Wolf to hand them the weapons.

We did as planned. I opened the station door and shouted, 'Hands up.' The policeman on duty did as I commanded. 'Where are your weapons?' He showed me the cabinet. I tried to open it but it was locked. I asked for the key and the policeman went to open a drawer in his desk. I shouted at him, 'Keep your hands up.' I went to the desk and opened the drawer, took the key and saw there was no pistol. I opened the cupboard and asked Wolf to take out the weapons and ammunition. I told the policeman that if he wanted to stay alive, not to phone the Germans before morning. He could say that the partisans had come by truck and taken the weapons.

We took the rifles and walked back to the forest. The boys were impressed at how well I knew my way around the forest at night. Marian turned up two days later and congratulated us on the action. Everyone now had his own weapon and there were two rifles to spare. Marian had already heard what had happened from the police. He was friendly with them. They said there were about 20 partisans; they had arrived by truck and the duty officer had no chance.

Marian said in a few more days two more boys would be coming from Piotrkow. We should carry on with our exercises. Everyone should be responsible for his own rifle. Marian returned two days later. When he got back, I lined up the four partisans in a row with their rifles. I wanted to show him how well disciplined we were and that we could all now use weapons.

We were told that we would have to carry out more actions. After the Germans started retreating from the Russian front, many wounded soldiers had been brought to the Tomaszow hospitals. The German authorities – fearing partisan attacks – moved German ethnic farmers, scattered over different villages, into one village, Pruchinsko, moving out resident Polish farmers to the Germans' former homes. Pruchinsko occupied a strategic position on the main road from Piotrkow from the west and Opoczno to the east, and was protected by the German Army.

Our task would be to fight the Germans as well as collaborators. A group our size would be more suited to that kind of operation. After much discussion about our duties, Marian said, 'Now, I would like you to take the oath for the Armia Krajowa and to our country. Our group will be known as Meva ['seagull'].' After the ceremony, Marian shook everyone's hand, congratulating them on now being a soldier in the AK. We had insignia in the shape of a seagull made by Stefania, a friend of Marian's, which we sewed onto our breast pockets.

I asked the boys, 'Which one of you is good at cooking?' Wolf replied that his brother, Zbik, was a good cook. So Zbik was put in charge of cooking the food. The boys then went to their bunker and started to clean their rifles and practise their aim.

Marian and I sat outside my bunker for a discussion about the group. He said he had found out that after our attack on the police station, two plain clothes Gestapo officers had come to the station to interview the policeman who had been on duty. After that, the police received instructions that no further weapons would be issued to them.

'Now, Florian, you have stopped being a full-time hermit, and have become a full-time partisan. I want you to come to my place so I can introduce you to my friends in the underground.'

Marian's German neighbours had moved from their farm to Pruchinsko, though Rudi came quite often to check on the farm where a Polish farmer from Pruchinsko was now living. He would always pop in to say hello to Marian and Janina.

I asked Marian, 'If I come to your home, and Rudi's there, will I be safe?'

'Don't worry,' he replied. 'He knows that you are in the forest with the partisans. He would rather help us than the Germans.'

A couple of days later, Marian came to the forest and brought two more boys with him. Andrey and Tarzan were from Piotrkow , both in their twenties. Andrey was very tall and intelligent. His father was killed in 1939 fighting the Germans and his mother worked in a shop in Piotrkow. He had an older sister who stayed with their mother. Tarzan was very strong and well built, hence his nickname; not too clever but very mischievous. He had left his mother, father and three brothers in Piotrkow.

Both had to take the oath straight away before their training so that the group could feel united. They stayed with the group in the second bunker.

I was to go to Marian's home before sunset and from there we would meet his friends. The first person we called on was Anton Zaborowski, whose small farm we had visited before the group's first action. He would be a contact if we needed one. He was short and skinny and had a great sense of humour, which he shared with his wife, Bronka. She was fat, bigger than her husband and very energetic, the mother of two little girls. She was surprised I was a partisan as I was so young. 'You can stay with us. I could do with a boy like you round here.'

From there we went to a farm belonging to the Mijas family in a neighbouring village. They had a daughter and four sons. The boys were all working for an underground group belonging to the AK. The youngest boy, Jozef, was in the Maja group; one of his comrades, Janek Skladowski, was with him. Janek was also a friend of Marian's. He was the son of a poor widow who had a smallholding near Blogie.

The Maja group was one of several partisan groups active in the area belonging to the AK. They were well equipped with supplies from Britain.

The Germans were very active in the area. From time to time, early in the morning, they would surround a village and arrest farmers. They had a list of the people they wanted, about 20 or 30 men from each village. Most of the farmers arrested were not involved in the underground. They were all taken to Tomaszow where they were interrogated. The underground tried to find out what was behind the arrests.

One afternoon in April, Marian came to the forest and told me that the following morning, German farmers were to go by horse and cart to Tomaszow to take gifts for the wounded German soldiers in hospital.

Marian heard about it from Rudi, who knew that Marian would pass this news on to us so that we could stop them and take the parcels.

We waited and hid in the forest near the road to Tomaszow, where we stopped two German farmers. We took the load from the cart and told them to go back home, and not tell anyone what had happened.

There was too much food for us to eat so we gave some cake, chocolate, sausages and ham to some of the poor farmers that we visited, and some others who Marian thought were sympathetic. I felt like Robin Hood.

Collaborators and Traitors

In a small village near Blogie was a German farmer named Zinger. Before having to leave his farm for Pruchinsko, the farmer went to check on his fields adjoining the forest. He wandered in to the forest and ran into a partisan group.They wanted to hold him for a few days until moving on. The farmer tried to escape and was shot. The partisans buried him in the forest, and he was reported missing by his son. The Germans came to Blogie with some farmers who helped them look for Zinger in the forest. There they found his grave.

Zinger's son had been in the Polish Army before the War. He was in his thirties, tall and stocky. He began taking revenge for his father's death, terrorising people he thought might be in contact with the underground or partisans. He carried a stick and beat up anyone he took a dislike to, mostly young men. He hurt a lot of people and warned that anyone who collaborated with the partisans would be arrested.

The underground sentenced him to death for his threats and actions. It was too dangerous to carry out the sentence because he only moved about during daylight hours and the village he lived in, Pruchinsko, was defended by German soldiers.

Some days later Janek, one of the four Mijas brothers in the Maja group, came to the forest by bicycle from Marian's and said to me, 'Florek, take the bike and go to Blogie. Zinger is in the village. Marian will wait for you by the church.' Janek told me that Zinger had bumped into him in Blogie and barked at him to bring him a bottle of vodka at the butter factory manager's flat in the village. Janek protested he did not have any vodka. Zinger shouted, 'You know where to get it.' Janek said he would try and then reported to Marian before coming to me.

I took the bike and went into Blogie. Near the church, I met Marian talking to a policeman. I stopped and said to the policeman, 'What's going on in the village?' He said that it was all quiet. I pretended not to know Marian and said, 'Who are you?' I was in uniform and they could see I was armed and a partisan. I told Marian to take my bicycle and leave it with the mayor; I would collect it later. I told the policeman to go to the station, to stay there and not make any phone calls. He agreed and went off.

After the policeman had gone, Marian told me that Zinger was at the manager's flat and drinking vodka; I went to the house and opened the door. Holding my pistol I shouted 'Hands up!' The three men present put up their hands. I asked their names. When I came to the name Zinger, I told him to lay face down on the floor and he obeyed. I searched for a weapon but there was none. All he had was a baton next to his chair. At that point Wolf arrived. He had been in the village and had come across Marian. We marched Zinger off to the forest holding his hands on his head. As we passed the police station we saw them watching from the first floor window, but they took no action as they had no weapons.

Marian informed head office and they sent an interrogator. Over the next few hours, Zinger said he was sorry for what he had done; it was because of his father's death. From now on, he would do what we asked if we let him live. We did not believe him; he could be more of a danger, and the sentence was carried out in the forest by firing squad.

The first execution squad had two volunteers, Tarzan and Wolf, the third person was chosen by Marian. After the execution, the boys discussed how they felt. It was their first experience. No one could be sure whose bullet was responsible for the fatal shot.

After several weeks, the regional command of the AK discovered that one of the arrested Polish farmers had managed to escape. He told our interrogator that the Gestapo showed him a raffle ticket stub that proved he had given money towards buying weapons for the partisans. He could not deny it, but he explained that he lent the money to help the man because he was unemployed and had a large family.

The farmer said a stranger, aged around 40, was collecting

money, saying it was to buy weapons for the partisans; he gave out tickets, saying they could get a refund with interest after the War: those farmers who donated money were all arrested by the Gestapo.

One evening, the man collecting the money came to the home of a young woman who acted as a liaison for the underground. He wanted 500 zlotys to keep his mouth shut. She denied having any connection with the underground and said she did not have that sort of money. But she agreed to borrow it to avoid arrest. He gave her three days to come up with the money.

Later that evening, she related what had happened to her underground contacts. Marian was put in charge of organizing the action. I took my three boys and we waited for three days, from daybreak to late into the night. There was no sign of him. Early on the evening of the fourth day, we spotted him arriving on his bicycle and entering the woman's house.

A short while later she came out and went to her neighbours, where Marian was waiting. Marian then came out of the house to the barn where we were hiding and told us the man we were after was in the house and we should wait by the door and grab him when he came out.

The woman told the stranger she had given the money back to her neighbours after he had not shown up. He fell for it and told her to go to the neighbour and get the money and to get back in five minutes. When she did not show up he came out to see what was happening. We arrested him, tied his hands and took him to the forest for interrogation. In his pocket was a book of tickets with 56 names of farmers who had given him money.

He confessed that he had started working for the Gestapo after they arrested him. They had accused him of being an organizer for the underground and said they would let him go only if he agreed to co-operate and pass on information. He came up with the idea of pretending to collect money for the underground. He kept the money and told the Germans which farmers were sympathetic to the partisans. His excuse was that he had a large family to support. He was an alcoholic; he said he could not live without vodka. We found a bottle in his bag. He said he was sorry and was prepared to take responsibility for his actions.

The interrogation was carried out by a delegate from the AK regional command. The stranger was sentenced to death and shot by firing squad. The interrogator took the list of names of contributors with him. The stranger's dirty work proved useful: the underground contacted the farmers on the list and explained how they could help the real underground.

Marian told me that if I needed a supply of food I could go to the vicarage in Blogie. They were sympathetic to the underground and would help out. One evening, I went to the vicarage with my two boys and asked the priest if he had any food for us. I explained that we were a partisan group and occasionally needed help from the villagers. The priest called his housekeeper, Tosia, and told her she should give us food whenever we needed it.

The housekeeper was a woman in her late twenties who called the priest 'Uncle.' I became friendly with her. She was an attractive woman and quite often when I came into Blogie I visited her. She always gave me food and drink and on Saturdays she usually made fresh vanilla ice-cream, which I was very fond of.

The vicar would often come into the kitchen while I was visiting, kiss my head and say, 'When are you going to free us from the Germans?' I told him, 'We're doing our best. Our day will come.' He told us to bring some flour before the holidays and Tosia would bake us a cake.

Over a year or two, Tosia and I became quite close, but not too close. We kissed and cuddled. She wanted us to have an affair but I declined. She didn't know I was Jewish and I didn't want things to become complicated.

In June 1943, a column of German soldiers stopped in a village near Blogie on their way to the Russian front. Two German officers came in to Blogie by horse and cart. There were some partisans from the AK in the area in civilian clothes. They grabbed the officers, took them into the forest and killed them.

The next day a group of German officers came by car and told the police and the mayor that if the two missing officers were not freed by the following day they would burn down the village. The mayor knew people who were connected with the underground and passed on the Germans' warning.

All regional partisan groups were ordered into Blogie and prepared to protect the village. The Germans in Strzelce, a village on the main road to Opoczno, loaded their lorries, but instead of coming to sort out Blogie carried on on their way to the front. The villagers had been prepared to run away, expecting German reprisals; instead, we were unexpectedly left in peace. It was another indication that the Germans were starting to lose the War. The partisans occupied the village for two days and then went away.

Instead of staying with my group in the forest all the time, I spent more time in the villages. It meant I could more easily stay in contact with Marian, make contacts in the villages and receive food from farmers.

One day, I was on my way from the forest to see Marian. We bumped into each other near Blogie. He was walking with a young woman, Stefania Wyzykowska, before coming to meet me. Marian introduced me as Florek – the partisan group leader of Meva. After we went in to the village together, Stefania left us and asked if we would like to visit her sometime. I went back with Marian to his home where we planned our next move, to punish or warn people who had contact with the Germans.

The next day, I went with my group to stay at a farm in a village not far from Paradyz. The plan was for me to go to the vicarage in Paradyz, contact the priest and collect information on collaborators and sympathizers. It occurred to me that Marian should be doing this job. It was far safer for him. If he was stopped he had papers. If I was stopped all I had was my pistol.

Marian was a good organizer and commander and co-ordinator between our group and regional command, but he wasn't brave or decisive. When he went with us on an action he would rather stay in the background, as he did with Zinger. He was right to be careful not to be recognised, but there were times when he could have taken direct action. He joined in interrogations and executions, but they posed no threat to his safety.

I didn't even think about the risk of getting killed because I wanted to fight so much and take revenge. I didn't stop thinking about my family; they were always in the back of my mind. By liquidating collaborators, I reasoned, I was saving people from the concentration camps and certain death.

At the farm I met Sophie, a girl from Warsaw who was staying there. The farmer allowed my group to sleep in his barn. I planned to stay in the area for a few days. I grew friendly with Sophie, a bright and pretty 20-year-old with blonde curly hair. I asked her, 'Why are you staying here when you're a Warsaw girl?"

She explained, 'I was afraid the Germans would take me as forced labour, so I got married to a friend. But he was arrested a week after our wedding. He was involved with the underground and I was frightened they would come for me next.' She had got the address of the farm from a friend in Warsaw who had relatives in the village. 'While I'm staying here I'm helping out on the farm,' she said.

She asked me if I had a girlfriend. I told her I had no time while I was fighting in the partisans. What with moving from place to place, I didn't even know if I would survive the War; it didn't make sense to have a relationship.

As the week-long friendship progressed I felt she was falling in love, but I could not respond. It was too risky for me. We cuddled and petted but I could not sleep with her and reveal that I was Jewish.

'I'd like to go with you and fight with the partisans.'

I told her, 'You wouldn't like the life we lead. It's uncomfortable and very dangerous. You'd be better off staying here until the end of the War. Whenever I get the chance I'll visit you.' A few Polish partisan groups had women, but they were rarely fighters; they were usually the girlfriends of commanders. I felt Sophie had too romantic a notion of what it was like in the partisans.

On the day I had to rendezvous with the priest, I asked her to keep me company on my way to Paradyz. A couple would look less conspicuous. She accepted and we went together. Near the church was a park with benches where I asked her to wait for me. I went to meet the priest in the vicarage. The housekeeper opened the door and I asked for the priest by name. She took me to his office. I felt slightly uncomfortable in front of the large cross on the wall and religious icons. I thought how strange it was for a Jew to be in this situation.

When the priest came in we got down to business: there was no talk of religion. I gave him a password. Passwords were given out by the regional command and passed to the two sides due to meet. They were only valid for one meeting. It might be something seemingly innocuous like, 'Have you finished that book?' to which he might reply, 'I am still reading it.'

Not all priests were sympathetic. One priest in Piotrkow, who gave confession to the Gestapo, was shot by an AK group. This priest was in his thirties, slim, intelligent, good-looking and very friendly. He gave me information about certain people who were in sympathy with the Germans and asked me to join him for breakfast.

Sophie was still waiting for me in the park, and we walked back together. The bacon at the vicarage had been very salty and I suggested stopping for a drink at a bar on the corner of the main road. Sophie declined; women were not made welcome there. She said she would go slowly and I could catch up with her after finishing my drink.

I was taken aback when I went into the bar. It was empty except for the barman and a German sergeant sitting at a table with a glass of beer and several empty glasses. I asked the barman if there were many Germans in the village. He answered that the one at the table had been sitting there alone drinking for hours. He thought he was waiting for someone or he had got lost. I paid for my beer and went over to the German. I took out my pistol and told him to put up his hands. He was drunk and unable to stand straight. He was short and fat with a red face and had to lean against the wall. He was silent as I took his pistol from his case and left the bar.

Running to catch up with Sophie I showed her my booty. She was very frightened. I said to her, 'How can you join the partisans if you are so frightened of this?' We carried on, keeping a careful eye out, before arriving back at the farm.

There I planned our action. We moved from village to village. We made a distinction between sympathizers and collaborators. Sympathizers were warned if they carried on and became collaborators they would be shot. After two days our work was complete.

I said goodbye to Sophie, who was crying in my arms. I told her if I got the chance I would visit her, but I never did. I returned with the group to our forest home.

Marian had a new task for our group. In a village near Zarnow lived a farmer by the name of Baranski, whose son was a Gestapo collaborator. He gave away some activists from the underground, who had been arrested and sent to concentration camps. He liked to terrorize people, robbing many farmers of money if he heard they had sold an animal. The AK sentenced him to death.

A group came to his village to execute him. He was not home when they called, giving his mother time to alert him. He went into hiding, though that did not stop him carrying on his dirty work.

The group came again and again without luck. In frustration, they took his father hostage and told the family the son had to give himself up before his father would be released. It made no difference. The father was held a few weeks before being shot and killed trying to escape.

There were individual members of the underground in the village watching for Baranski. They did not keep weapons, and anyway could not risk the lives of their families by executing him. After a time, things calmed down and Baranski started visiting his mother, but he was insecure, watching everyone and everything around him.

In September 1943, Marian had orders from head office to use our group to get the traitor. We decided that the three boys and I would go to the village on the following Saturday and meet two young men from the underground who lived there. As we approached a small wood near the village, they stepped out and asked for the password. I gave it to them. We sat down in the forest to discuss the situation, waiting till it got dark. They took us to the village to show us the farm. I went with one of my boys into the house while the other boys secured the back. We looked everywhere but could not find him.

We found a new bicycle hidden in the straw. We gave it to

the boys from the village. They would find out who it belonged to and give it back to the owner. I did not want to go back to Blogie without finishing the job. I decided that I would stay in the village with one of my boys and the other two should go back and tell Marian what was happening.

The underground boys took us to a farm where the farmer's young adult daughter was a member of the underground. Her parents did not know about her double life and she hid us in a barn without their knowledge. She prepared a place for us to sleep and brought us food. In the morning, she brought hot milk with bread and butter. She told us that she would go to the village to meet some friends and find out what she could about Baranski.

On her return, she told us she had seen our traitor standing outside his home. He had asked her to go with him to Blogie to the annual Holy Mother's Day festival. She declined, saying, 'There are so many girls in the village you could take, why are you asking me?' and went off to her friends.

I was ready to go and execute him there and then, but the girl warned it was too risky to do it by daylight. The Germans could punish the whole village. If he went to Blogie it would be better to get him there.

I asked, 'How will I recognise him?'

'He's not tall and has a round face full of spots. He's wearing a brown suit and a white shirt with regional flowery stitching.'

I thanked her for her help. We went on the road to Blogie. On the way, we got a lift from farmers going to the festival. When we arrived in Blogie I went straight to Marian's house and told him of our plan. We had lunch and I changed my clothes. I put my pistol under my belt, covering it with my jacket; then went with Wolf to the festival.

We looked for our man as we walked through the crowd. Then I saw him, sitting on a wall between two men. He matched the description of the man we wanted. Wolf told me he had to leave, as he had seen several Germans from Piotrkow and they might recognise him. I asked him to go to Marian to check if I had spotted the right man.

While Wolf was away, the three men came off the wall and walked into the crowd. I got a good look and realised it must be him. I came from behind the three men and, pressing my pistol into the back of the one in the middle, told him to walk straight ahead. I ordered the other two to move away.

We got behind a fence away from the crowd. 'Hands up,' I commanded, and asked his name.

'Baranski.'

It was the name I was looking for. I checked his pockets for weapons but found nothing. In the meantime, Zbik came along. I told him to watch Baranski and to keep his hands up. Seeing a cart and horses nearby I asked the farmers standing around it, 'Whose horses are these?' One of them answered, 'They are mine.' I showed him my gun and said we were taking the horses. He would have them back in an hour.

I told Baranski to lay face down on the cart. Zbik covered him with blankets. I drove the horses through the crowd shouting, 'Get out of the way.' A path was quickly cleared.

When I reached the forest the boys took hold of Baranski. Zbik took the horses back to the owner and Marian sent an interrogator. Baranski told us that a friend of his who worked for a German farmer earned a lot of money working for the Gestapo. His friend introduced him to the Gestapo and he agreed to work for them, too. He was sorry for what he had done, but after his father was taken hostage by the AK and shot trying to escape he could not give up. After having written down all he had to say, a messenger went to the regional command based in a forester's house just two kilometres from Blogie. He came back with a written sentence of death. We formed a firing squad and executed him that evening. His friend was executed by another group.

Several weeks later a woman, aged about 40, well educated and comfortably off, moved into the village where Baranski had lived. She told farmers that she had had to move away from Warsaw because she was frightened of the Germans, and in the village she could live without fear. She was too friendly and there seemed to be no reason for her arrival. Most strangers who turned up were either driven out of their homes by the Germans or were members of the underground on the run.

The underground was wary. The Gestapo was bound to have been suspicious about Baranski's disappearance. A member of the underground who lived in the village, and who was originally from Warsaw, asked her what part of Warsaw she came from. It was clear she did not know the city.

The underground kept an eye on her and spotted her meeting a man in a car outside the village. She tried to join the underground without success; no one admitted it existed in the village. A member of the underground from outside the village posed as a fellow collaborator and she confided in him, admitting she worked for the Gestapo. She had not come from Warsaw but from Opoczno and had been sent to the village by the Gestapo. The supposed collaborator told her about Baranski, who she had not known about before her arrival.

When the underground were informed of what she had said they sent our group to the house she rented in the village. We told her she had been sentenced to death and one of our group shot her in the head. The mayor was informed and told to take care of the body.

The temptation of having weapons and power proved too much for some partisans, who turned from fighting the Nazis to robbing and murdering our own people. Home Army command ordered Marian to use our group to track down just such a former AK partisan- turned-bandit.

Karol's group of eight to ten men were based around Tomaszow, which was normally outside our area of operations. They had shot a few of their own comrades before setting about their criminal activities, robbing people in their homes, taking money, jewellery, food and drink.

Word of their crimes reached the Germans, who somehow managed to locate Karol, saying they wanted to make him an offer. They would turn a blind eye to his crimes if he carried on attacking the partisans. In return, they would arm his group. He agreed, and they supplied him with rifles, ammunition and money.

The local Home Army command had sent a couple of partisan groups after him, but without success. We were then brought in. We spent a few days in the area but could get no leads. Then we went to visit his wife in the house that he had newly built. We searched the house, the cellars and the loft, but found nothing. His wife said she hadn't seen him for a long time, didn't know where he was and didn't want to know.

We searched the stable and the barn. At first we found nothing. Then I noticed something odd. The barn was stacked full of straw – but there was a little gap between the straw and the wall running the length of one wall and only big enough for someone to squeeze through sideways. I shuffled through to the end and found a half door, which I opened, leading to a bigger gap in the straw. I saw some bedding and shouted, 'Put your arms forward and move out.'

Karol was there. He had been asleep and was still lying on the floor. He put out his arms and crawled out. Once through the gap, he stood up. I held his arms and led him out.

He said he had hoped to hide away until the end of the War. He was alone and unarmed and made no attempt at resistance. We were unable to find his comrades or his weapons.

Marian had orders to execute him immediately if found. Usually it was by firing squad or a bullet in the head. This time the order was for him to be hanged. It would set a better example for any others thinking of turning to crime, and show the Germans that the partisans were a disciplined and principled force.

Taking a rope from a nearby farmer the boys strung him up from the branch of a tree. The cart was pulled away and as he dropped, Marian ordered one of the group to shoot him in the head, just to make sure.

We had been ordered to stick a notice on his dangling body. 'I was a collaborator. My body should be left to hang for three days.'

When the Germans found him, an officer ordered that he should be left there like the note said. He told the mayor they would like to give his killers a reward of 10,000 zlotys as Karol had betrayed the Germans too by not honouring his side of their agreement.

For some weeks, our group stayed in the forest and rested. In that time, a large group of partisans from the AK put up near our camp. I spoke to their commander and was surprised that he knew of our group, Meva's, existence.

Next morning, the commander asked me where we got our butter from; I told him from the factory in Blogie. I said, 'I'm going to Blogie today.' He asked if I would mind taking his sergeant with the horses and cart and would bring them back some butter, too. 'If you know of a shoemaker please bring him back, and he can repair shoes for the boys.'

I did as he asked. First I went to the butter factory, and when I went in and asked for the manager a worker told me that two German engineers building an airfield near Pruchinsko were also waiting for butter. I went to the office and with my pistol pointing at them, told them to put up their hands. I asked if they had any weapons. They answered, 'We are engineers, not soldiers.' They told me they were there to collect a ration of butter.

I said, 'Today we are taking the butter.'

'Good. Good,' they said. 'We will come again tomorrow.'

I told them that if they wanted to stay alive, not to tell anyone about our meeting. They understood a little Polish.

We loaded a box of butter on the cart and went to another village, Blogie Rzadowe. We arrived at the shoemaker's. His wife told us that he had gone to Sulejow to buy some leather and would be back for lunch.

There was a friendly farmer nearby, so I went to visit him, together with the sergeant and Wolf (who was on his bike). The farmer's daughter cooked some scrambled eggs. The farmer put a bottle of vodka on the table and told us to eat and drink while we waited for the shoemaker to return. I asked Wolf to cycle to the forest and tell the group leader what was happening.

About lunchtime, the shoemaker arrived. He ate his lunch then took his tools and leather and came with us. We drove into the forest not far from where the AK group was camping. I caught sight of an army hat that I had given to Wolf. It was laying on the track. I jumped off the cart and picked up the hat. I saw it was torn and realised that something had happened, but the sergeant decided that we should go on to the camp.

We arrived at the last crossing before their camp and from a distance could see a group of soldiers in green uniforms standing where the partisan camp guard had been. We waved to them and they waved back. At the end of the path was another group of soldiers. I jumped off the cart and went looking for someone from our group. All I could see were a lot of shells from machine guns. Something was wrong. As we turned around and drove away, the soldiers at the other path started to shout, but we smacked the horses and got away, driving off the track and heading for the forest ranger's house. I stopped and went to the back fence. As I approached, the dog started barking. The door opened and the forest ranger came out and looked around. I gave a whistle and he came to the fence.

'What's happened to my group and the other partisans?' I asked. He knew me – Marian had introduced us – and knew about us staying in the forest.

'You don't know what's happened?' he asked.

'No. I have been in Blogie for butter and we were waiting for the shoemaker.'

He told me that in the morning, a group of German and Hungarian soldiers came to one of his rangers. A German officer showed him a map and asked him to take them to a point in the forest. They attacked the group and took prisoners, three Russians and two Poles. One of them was Wolf. The rest of the group managed to escape and dispersed.

I went to Marian and told him what had happened. He knew already from another contact. We decided that I should wait at Zaborowski's farm near Blogie.

The large AK group, about a hundred of them, reunited that night in Blogie before moving out of the area. The next morning, while staying at Zaborowski's, I saw Wolf dart out from the edge of the forest. He had no shoes and the bottoms of his trousers were torn to pieces. We let him rest, gave him some food and drink, then he told us what had happened.

After leaving me waiting for the shoemaker he returned to the forest. The Hungarian soldiers stopped him on his bike. He thought at first they were Polish partisans as they had the same colour uniforms. They took him prisoner. The group had been attacked before he had come to the forest. He and the other partisans were held in a village where Hungarian, Czech and German soldiers were staying. Three of the AK partisans were Russian. The German officers took them away to Tomaszow.

'How did you manage to run away?' I asked Wolf.

'A friendly Czech officer told me to go and bring some water with a German soldier as guard, and then to run away. I went to the well and pulled up a bucket of water. The German stood below the mound of the well with his rifle on his back. I put the bucket full of water on the German's head and ran over the field into the bushes, and then found my way back to Blogie. The German fired, but I was away by then.'

'Good on you,' I said.

I found out later that the Czechs helped the remaining Polish partisan to escape. For the partisans there was a problem. How did the Germans find out about the group's camp?

The local underground organization had to find out who was likely to be giving information to the Germans. One of the partisans, who had been on guard a few days before, reported seeing a well-dressed woman and a man; he thought they were holidaymakers. They were collecting mushrooms in the bigger forest nearby. He was not sure if they had seen him, but he was sure they could hear noises from the partisans. I decided to move us away from the forest for a while. Marian agreed.

Seven of us went on the vicarage horse and cart driven by the priest's farm worker, who dropped us off in Dabrowa, 15 kilometres away. We had no food or utensils; everything had been confiscated. I went with the group to a large farm we knew of where we got something to eat, though they did not have anything for us to take for our journey. They said we could get food and a cauldron for cooking from the local vicarage.

We arrived at the vicarage at midnight. The housekeeper was in the kitchen. She let us in and said the priest was asleep. I told her we wanted to buy some bread and meat and needed a cauldron or pan. She went to wake the priest, who came to see us. I repeated our request, and explained that the Germans had attacked our group and we had left everything behind. The priest said that he had nothing because another group of partisans had taken everything.

'Haven't you any chickens or geese?'

'No, we have nothing, we only keep going with the help of locals.'

I asked for a horse and cart to take us to the forest. His worker could go with us and then drive the horses back. The priest called the farm worker, who was asleep in a little room adjoining the stable. He came out and started to prepare for the journey. One of my boys asked him, 'Was the priest so poor that he did not have bread, or chickens?"

The man told him that that very day, the housekeeper had baked fresh bread and there were plenty of chickens and geese in the stable that he showed him. I asked the priest to open up the locked stable. It was full of birds. I asked two of our boys to take a couple of geese and went to the kitchen to ask the housekeeper where the bread was kept. She showed me the door to the pantry. It was locked and I told her to open it. When the door opened, the smell of fresh bread hit me. I called one of the boys and told him to take a loaf. I turned to the priest, 'Do you think you are carrying out God's wishes?' He remained expressionless and said nothing.

As we prepared to drive away, he came over to me and said, 'Aren't you going to pay for what you've taken?'

I replied, 'We asked if we could buy food from you but you denied having any. I'll let God repay you for your kindness.'

The labourer took us to a neighbouring forest where we got off the cart, walking the rest of the way through the forest to our destination: a farmhouse at the other end. We informed the farmer of our presence and slept on straw in his barn. We rested for two days before going to my old hiding places around Siucice. I felt safe there, and the farmers were pleased to see me, and my group. Every few days we would move around the villages for safety.

After a couple of weeks we went back to Blogie. I left the boys at Anton's farm, on the corner of the road that led to the forest. I went on to Marian's house. Marian and Janina were home and greeted me with their usual warmth. After a drink and some food, Marian asked me to come into the other room. There I told him what had happened. The most surprising thing for him was the way the priest at Dabrowa had behaved towards us.

Marian told me that things had been quiet for the last two weeks. The organization found out that the two people who were collecting mushrooms were teachers from Tomaszow, staying in the village behind the forest. They claimed to be on holiday, were renting a room at the farmhouse, and were not short of money. They were very active and moved about a lot. Sometimes, visitors from Tomaszow came to see them by car. They had been seen meeting someone outside the village.

Marian said that in a couple of days, he would have a decision from the underground on whether they should be interrogated. I returned with my group to our forest. The boys were more comfortable there, enjoying the freedom to move around.

Late one afternoon, I went to see Marian to check on the situation. On the way out, by the main track not far from our bunker, I heard noises behind the bushes. I stopped and reached for my pistol. Behind the bushes I saw five men sitting on the ground.

'Who are you?' I asked.

'Mey Ruskie Plenniye' [We are Russian prisoners], came the reply.

I saw from their mish-mash of clothing that they were poor and had been working in the fields. Their shoes were full of mud.

I understood Russian and they explained that they had run away from Tomaszow and wanted to join the Russian partisans. I explained to them that in this area there were no Russian partisans. I took them to my group and asked the boys to look after them, give them some food and watch them carefully until I came back.

Marian was not at home when I arrived. Janina told me he had gone to visit a friend and would be back for supper. As I was having some coffee, someone knocked at the door. I went into the bedroom and Janina opened the door. She called out, 'Florian. Come out. It's Franek.'

It was Franek Mijas, a friend of theirs. I came out with my coffee and chatted to him. We had met before. He knew about our group in the forest. One of his brothers, Jozef, was in the Maja partisan group, together with Janek Skladowski. When Marian got back, he asked me to come to the other room. I told him about the Russian prisoners. I suggested we keep them in our group. We had enough weapons and they could be useful for fighting.

'Keep them,' he said. 'And watch them until tomorrow. I will come to the forest and tell you what to do. By then I will have a decision about the holidaymakers and the Russians.'

I returned to the forest that evening to check that everything was alright. The Russians were resting in the bunker and the boys were outside by the fire. They had their weapons with them and the spare ones were in my bunker.

The next afternoon Marian came to the forest. I introduced him to the Russian prisoners. He told me they could stay in our group. We would give them rifles after our next action, due the following evening.

Marian returned in 24 hours to go through our plan. We would go to the village where the holidaymakers were staying. Two men from the underground would be waiting for us. They would show us which house the couple was staying in. We would surround the house; then I was to go with Zbik and Wolf and bring them out.

After sunset, we left our camp and went to the village behind the forest. Two men were waiting for us. We exchanged the password and sat down to finalize the plan. It was Friday night. The men knew the couple was home.

Come midnight, we moved into the village and went directly to the house. We knocked on the door; it was open and we entered. The farmer got up from his bed and lit the paraffin lamp.

I asked him, 'Which room are the holidaymakers in?"

He showed me the door and said they were asleep. I opened the door; the light was on. The man was putting his trousers on. I ordered him to put his hands up. The boys searched him and the room. They found no weapon, but many things of German origin, such as food and medicine. I told him that, in the name of the Polish underground, he was being arrested. The woman, who was still in the bed, started shouting.

'Where are you taking my husband? I won't let him go, I can't stay here alone.'

When he was dressed and ready, I asked her to get up and

put on her clothes. She said, 'I am not going with you. I'll stay here alone.' I asked her once again, this time she obeyed.

We took them to our camp and told them they could rest and sleep in the large bunker. In the morning, Marian began the interrogation and was joined later on by an interrogator from the regional command. He recorded the proceedings on paper.

The couple confessed to working for the Gestapo. They had worked as teachers at a school in Tomaszow, where they were arrested and accused of belonging to the underground: a routine accusation. They had to either agree to work for the Germans or get sent to a concentration camp. They started spying in Tomaszow. Then the Gestapo asked them to go on holiday to a village adjoining the forest and find out who was working for the underground. They did not find out anything, as the villagers did not trust strangers. When they went to the forest to collect mushrooms they came across our camp that temporarily included the large AK group. They said they had not believed the Germans would go in to the forest to fight the partisans and apologised for what they had done.

From their documents, we could see that they were not married. The man confessed that he had a wife and a four-year-old daughter in Tomaszow. The Gestapo thought it would be better if they posed as a married couple.

Their confessions were taken to head office and on the same day, a sentence of death was returned. The delegate from the office read them their sentences. The woman stayed silent, her head bowed. The man accepted the decision, saying, 'I deserve it. I betrayed my country and my family.'

He was asked if he had a last request. 'Please let my daughter grow up a good Polish patriot.' The woman said nothing.

He asked if he could have his jacket, because it would be cold in the grave. I handed it to him. He held it in his hand and said, 'No, no. I won't need it.' He put it down. 'Someone will find it useful.' He asked if he could face his executioners. We said no. They turned their backs and were shot.

A Scent of Betrayal

We temporarily moved from the camp to stay in the villages. We armed the Russians with rifles and they joined us in further actions. We were now a group of 12 partisans. We disarmed Germans in ones or twos as they came to the villages to grab what food they could (mostly eggs and chickens or vodka). We did not kill them unless they tried to fight us, as we did not want to bring trouble on the villages.

When a German company was stationed in a village on their way to the east, they would put guards at either end of the village while the troops slept on farms and in tents. On four or five occasions we came into the village at night. The guard would often be standing but asleep. We would come up behind them, put a hand over their mouth, a pistol in their back, and drag them off into the forest where we would shoot them. The Germans never sent anyone in search of them; they assumed they were deserters.

Marian had a new task for us. He had received orders for our group to execute someone similar to Zinger in a village near Opoczno. Usually, he left such operations to the group. But he had business with someone, so he would come along too.

Next evening, we went by horse and cart to Dabrowa and Rozynek, crossing over the River Czarna before joining a track that avoided the main roads. Some way on, the track split in two. Unsure which way to go, we stopped. I saw a light in a farmhouse window off the track and went to ask for directions accompanied by Misha, one of the Russians. I knocked at the door and without waiting for a reply went in. In the middle of the room sat a woman wearing a heavy jacket and a fur hat. A man was standing by the window. I asked him the way and he told me to take the first turning on the left. I asked the woman who she was but she did not answer, and neither did the man.

Suddenly, the door opened and a short fat man came in with a pistol in his hand and, in Russian, asked who we were. I told him we were Polish partisans. He said he was a Russian partisan. 'I am Major Gromov.'

Another man came in with an automatic pistol pointed at us, turning it away when he saw there was no problem. Misha was curious about his pistol and asked to see it. The newcomer refused. Misha said he could have his rifle while he let him look. The man took out the pistol's magazine and handed it to Misha. It was a new automatic that Misha had not seen before. He admired it and said, 'Good Russian weapon,' and handed it back.

I was not sure who the men were, so I said to Misha that we had to go. As we left, Gromov came behind us. Moving into the darkness I turned back, afraid he might open fire. Gromov called out to Misha to come back, he wanted a word with him. Misha asked me if that was OK. I told him to find out what he wanted. Gromov said they were a parachute group and asked Misha to join them as a fellow Russian. He said he would have to ask me, as he was a member of my group. When he came back over to me I asked if he was sure they were parachutists, but he did not know, as they were not wearing uniforms. I told him, 'You don't know and I don't know. Come on, let's go.'

We carried on on our journey and in the early hours of the morning arrived at a village close to Opoczno, where I had arranged to meet Marian. I went with him to meet his contact, who told us a large German Army movement was taking place; it was too risky to take our action.

When Marian and I got back to our group we found they had been disarmed by a fellow-AK group who were based in a nearby village; they were a huge group of partisans and their leader told Marian that he would have to wait till the next morning, as they were awaiting confirmation of our identity. It was not uncommon for different partisan groups to operate in the same area under the orders of the underground.

On this occasion, Marian was directly involved in the action. He had no choice; he was the one person in our group who was in contact with the regional commander.

That night, we stayed on a farm in the village and were given food by the partisans. Next morning we were handed back our weapons. One of the officers from the large group heard me talk to a Russian from my group. He took me aside and asked, 'You are still keeping them alive?'

I said, 'We are all fighting the same enemy.'

The officer laughed and said, 'We finished them off a long time ago.'

Then I found out that the AK partisans were with a group of NSZ (Narodowe Sily Zbrojne – 'National Armed Forces'), fascist sympathizers who fought the left-wing groups of the AL (Armia Ludowa – 'The People's Army') and who were pledged to rid Poland of all Russians, Jews and leftists.

We went on our way back to Blogie. In the forest, we saw from afar convoys of German Army lorries on the main road. After a while we heard machine gun fire. We were not sure if it was connected with the partisans or just German Army exercises.

It was not safe to leave the forest until dark, so we waited till evening before moving out, arriving back at Blogie late at night. Marian went to his home while my group stayed the night in a farm near the forest. Next morning, I went with two of my boys to the butter factory and collected a box of butter that the Germans were supposed to get. Then we went to the mayor of the village and asked him to get some bread for us, which we would collect in the evening.

We returned to our forest home. In the evening, Wolf and Tarzan went to get the bread and the two Russians went hunting, hoping to shoot a wild animal for our meal. They got a small wild boar and the boys brought two large loaves. We ate well and were able to rest. We smoked some of the meat and kept it in the rucksacks. The butter was buried in the ground in a barrel covered with a cloth with a thick layer of salt on top so that it would keep fresh for a long time.

After several days, Marian came and told us to prepare for an action. He had found out that a large group of German soldiers had moved away from the village of Bukowiec, leaving behind a small contingent of German, Hungarian and Czech soldiers. He suggested in the morning that we take two carts and, once close to the village, start making a lot of noise, shouting and firing weapons, so that it sounded like they were being attacked by a large group.

We carried out his plan and before we got to the village the terrified soldiers loaded what they could on to lorries and ran off. They left a lot of boxes with grenades and ammunition, dropping some into the well to destroy them. The farmers helped us retrieve them; it was all still usable. We checked each house to make sure all the soldiers had left. Then we loaded all the equipment on to the carts and went to the village community office, where we found two boxes of butter and Wolf's stolen bicycle.

Later, we stopped two German farmers on a horse and cart coming from Tomaszow. We locked them in a stable while we stayed in the village and told the other farmers to let them out after we had moved away.

Back in the forest, we unloaded our booty and stored it in the large bunker. We gave some of the butter to poor farmers and I took some to Stefania. She made me a woollen sweater for winter and washed my underwear and shirts; I no longer had time to do these sorts of domestic chores. A few farmers' wives were willing to wash my clothes, including those at the large manor farms where I stayed from time to time with my group.

There were still many manor farms in Poland owned by the rich, landed gentry. They were sympathetic to us and I would develop a routine when I went there. Their servants would provide food for the ten or twelve boys I had with me, while I would wash in warm water and change my underwear. It was important for me to stay hygienic and the landowners respected me for it. When I finished, I went to join the boys to eat. They washed themselves and their clothes in the river or on a farm; they were not too bothered about dirty clothes or personal hygiene. I would leave my dirty washing at the farms and would return to collect my nicely ironed shirts and clean underwear and leave a new batch to be washed. When there was no other possibility I asked Janina to do my washing. While I was alone in the forest I washed my clothes myself in the River Pilica. In the winter, I did it in the melted snow. Now I was facing winter again, no longer alone. I was with a group of partisans and had plenty of weapons.

We moved out of the forest to stay in the villages. I could have managed another winter on my own, although I would not have looked forward to it. It would have been impractical to stay there with a group. We needed to be mobile and could not have survived holed up in a bunker with the snow over our heads.

We went back to the forest from time to time, though not to our old bunkers. We spent the days in an area where the trees were larger and there was more room to roam. We made large fires that we slept around, warming one side and turning over to take the chill off the other. The winter of 1943/44 was very cold, sometimes going 30–35 degrees below zero. It was hard enough fighting the elements.

It was difficult to find out which villagers we could trust. Some of the villages supported the AK, some the left-wing AL. There were also partisan groups, such as BCH (Bataliony Chlopskie – 'The Peasant Battalion'), formed by farmers, some of whom were left wing, some right.

Most of the villages in the region where I stayed around Blogie were AK sympathizers. Villagers supplied us with food. They knew it was in their interests to co-operate. We stayed out of the villages some days as a security precaution; it was too dangerous to be in one spot all the time. At night we moved from village to village, resting a few nights in each place.

German troops did not move during the night and in the daytime they kept to the main roads. During the winter, all activities slowed down; we concentrated on survival. With spring, the snow rapidly disappeared as the sun rose brighter in the sky. I moved back with my group to our forest bunkers for safety. The bunkers had survived the winter well. They needed to be aired but, structurally, were still intact.

Marian asked me to visit him so we could plan further actions. The boys were getting bored moving from place to place without any purpose. Zbyshek fell ill and had to stay in Blogie with his uncle. I often went to the village to visit friends. One evening I went for a discussion with Marian. He was not home. Janina said he would be back late or maybe the following morning.

I went to the grocery shop, which doubled as a bar, and met friends, staying there for hours, drinking vodka and eating hot sausages. Late that evening we heard noises outside. I took out my pistol and opened the door, asking who was there. It was Jozef and Janek Skladowski from the Maja group of AK. Both came in and bought a bottle of vodka to take away. They refused our offer of a drink, saying they were on their way to the forest ranger's house. We thought nothing of it and carried on having a good time.

At midnight, Jozef's brother, Franek, came in and asked about Janek. I told him he had gone to the ranger's house. He said if Janek came back to tell him that two partisans were waiting at Stefania's home. He knew about what.

Before I left the bar, Janek and Jozef turned up again. I gave Janek the message from Franek and he left immediately, with Jozef following behind. I went on my way to Marian's but he was still not back. Janina made some coffee and as we chatted, she warned, 'Florek, you must be careful. Franek told me that the two partisans from Maja group were at Stefania's house asking a lot of questions about you. They want to know if Stefania knows your real name and where you come from and how she knows you. The two men must be friends of Jozef and Janek if they're in the same group.'

This was worrying news. I wondered why they wanted to know about me. Maybe Marian had let slip something about my origins while he was drinking with Janek? Or perhaps Janek was jealous that Stefania and her family were so fond of me? Whatever, I sensed I was in danger. At that time, life was cheap; people killed each other without a second thought.

I had a restless night and thought it would be a pity to be killed over nothing. To die fighting in battle was another matter. Being a leader of a partisan group was no protection if someone chose to take a pot shot at me. In the morning, I returned to the forest where I found the boys busy preparing food. Only Misha noticed there was something wrong. I put it down to too much to drink and not enough sleep. I carried on with my everyday routine.

The boys were starting to become disaffected; they wanted action and said they would prefer to join a larger group. On my visits to the village for information, or to contact Marian, I always took Misha as my guard. One evening, on our way to Blogie, we stopped off to see Anton Zaborowski. We had a drink and something to eat. Anton mentioned there was a group of Russian partisans at his brother's house at the end of the village near the forest.

I said to Misha, 'Come on, let's go and find out who they are.' I had been thinking about joining the Russians. I felt I would be safer with them.

When we arrived at the farm of Anton Zaborowski's brother, a Polish partisan ordered us to stop, pointing his machine pistol at us. I said, in Polish, that I wanted to see his commander. The guard knocked at the window, a man came out and we were led in.

There were two rooms full of partisans, some in Polish uniforms. The Russian group leader was Major Nikolei Gromov, who I had met the previous autumn. He did not remember meeting me. I asked whether I could join his group with my five Russians. He said, 'Speak to the Polish commander, Lieutenant Stefan, he's the leader of an Armia Ludowa group.' The Russians were under the protection of the Poles.

I introduced myself and said that I had five Russian ex-prisoners of war and that I would like to join his group with them because I believed we were all in danger. Stefan laughed, 'Does the AK want to put a spy into my group?'

I told him no one had sent me; I came of my own free will. I expected that, as a left-wing group, the AL were more inclined to protect Jews. I told him the story of the NSZ officer who boasted that they killed Russians, and that I knew partisans from the Maja group were collecting information about me. 'I am Jewish, and no one in the AK knew that except Marian Kowalski. He helped me hide and organizes the Meva group of partisans.'

Stefan asked my real name and where I was from. He told me to bring the Russian boys and leave all our weapons with the AK as he had no need of them. 'Hurry up, we're leaving tonight,' he said.

I went and told Marian what Janina had said about the partisans looking for me. He already knew; she had told him too. I said I felt in danger. I wanted to carry on being an active fighter; I needed to take more revenge. I had the chance to join a Russian group with my Russian boys and I had to take it.

I said the Polish boys could join a larger AK group, which is what they wanted anyway. Marian said he did not want to lose me. 'It's your life, your choice. I can't stop you. But think hard about what you are doing.'

I told him I had no choice; I wanted to fight Germans, not Poles. He knew I was right. He accepted I was in danger from Janek's AK friends. But he was concerned about me going to the Russians.

It all became very emotional. I thanked him and Janina for saving my life and said my goodbyes. I sent Misha to the forest to collect the five Russians.

I later learned that the regional command ordered Marian to disband the Meva group; the boys that were left joined the other AK group. Janina later told me that Marian carried on working for the underground, using his contacts to give them information.

Joining the Russian Partisans

That night, we moved with Gromov's Russian parachute group and Stefan's Poles to another region about 30 kilometres away. I had to get used to a new life as part of the Armia Ludowa. The Russian group consisted of 11 men, including two Germans, Karl and Werner. The AL comprised about 30 men in all; and myself, and my five Russians, brought us all up to nearly 50 men. That was quite a large partisan force and we needed a lot of space.

The Russian group had two radios. Gromov was the leader and his two lieutenants were Misha (this was another Misha) and Andrej. He was 20, slim and blonde, intelligent and kind natured. We were to become very close. There was also a young woman, Masha, a 21-year-old radio operator.

The Russian group kept in touch with their command centre in Kiev; we received all instructions from there. They had accurate information from their spies. Our principal job was to blow up German military transport on its way to the Russian front. The main railway line that the Russian group had to target ran from Tomaszow Mazowiecki to Opoczno and Kielce. The line was patrolled by the hated soldiers of the Wlasowa Armia (Ukrainians led by General Wlasow, who had defected to the Germans). The two professional explosive experts in the group, Lieutenant Andrej and Misha Stepaniuk (a third Misha) laid the solid blocks of 'Trotil' dynamite under the rails. The explosives detonated automatically when a locomotive passed over, by which time we were long gone.

I once offered to help Misha lay mines. He told me, 'A miner only makes one mistake in his life. We can't afford to lose you.'

A few days after an action, we would collect intelligence on how much damage we had inflicted. Our informants were nearby farmers taken by the Germans to clear the buckled tracks. They counted the coffins brought to take away the bodies. The information was passed on to Kiev. We carried out many such successful operations up and down 30 or 40 kilometres of track, working on information supplied over the radio from Kiev. They had excellent intelligence about the times of trains and we never mistakenly blew up a passenger train.

As I had been active in this region for over two years, I knew how to reach the tracks without maps, which was very helpful to the group; the commanders trusted my knowledge.

I became friendly with Karl and Werner, the two Germans, who told me they had expected to be parachuted into Austria, where their German language would be most useful. However, this was abandoned, as they were needed to replace a lost Russian partisan group. The Russians had been parachuted into our region but had gone missing, killed, it was believed, by a far-right Polish nationalist group. As the railway track they were meant to sabotage was an important supply route to the Russian front, the Russian command decided to send the group with the Germans to Poland instead to replace the missing group.

Karl and Werner were single men, both aged about 40. They were communists, and had fought in the Spanish Civil War in 1936 as volunteers for the Republican cause. After that, they went to England and then to Russia before the outbreak of war. In 1942, they volunteered to fight the Nazis. They both spoke reasonable Russian, which they had picked up while living there.

I found it hard to adapt to fighting alongside Germans, but I admired them for fighting the fascists. I trusted them completely and knew they were sincere. They were intellectuals and we had many discussions about Hitler and the German people. They taught me German and I taught them a little bit of Polish.

One evening, while our group was staying in the village of Alexandrow, a messenger came with orders for our commander, Stefan, to take his group and join an AL brigade on the western side of the River Pilica. We had our evening meal and some vodka. Stefan explained to Major Gromov that as he had to move, he would leave one of his partisans to help the Russians communicate with Polish farmers. Gromov said he wanted Lewko to stay with him. (Lew is Polish for lion; it was the name I took when I joined Stefan's group.) Stefan said to Gromov, 'Take anyone else, but not Lewko.' He valued my local knowledge and fighting skills. The Russian boys who had stayed with me in the forest begged me to stay with them.

Gromov and Stefan went into another room for a private discussion. Stefan came out and asked me if I minded staying behind because Gromov would not budge. Stefan said, 'He's right, if we leave them without any help they could end up sharing the same fate as the first parachute group.' I agreed to stay with the Russians, and said my goodbyes to the Polish group.

We stayed at the farm in Alexandrow where we planned our next move. Gromov and his two lieutenants, Misha and Andrej, wanted me to join them on their leadership committee. I said it would be an honour and accepted. That night, Gromov reported what had taken place to headquarters in Kiev and waited for further instructions.

The following night, we moved near to the main railway tracks. We carried out our instructions as before, taking the lines out of service for some time. My knowledge of the local villages and farms helped ensure we avoided coming across any hostile German or Polish groups.

After each action we moved location. One evening, in the autumn of 1944, I stopped with the group at a farm at the edge of the village of Dabrowa, close to a forest. We stayed overnight and through the next day. After our evening meal, a young man turned up: he was stopped outside by the guard, who knocked for me to come out. The man said he wanted to see the farmer. I let him in and then he started to speak to our boys, who did not understand Polish. He told me he had come from the next village. He tried to be friendly, trying to find out what we were doing there. He handed out cigarettes and we offered him a vodka. While he was chatting to the farmer, one of the Russians who came with me to the group called me by my old name, 'Florek,' and I answered him. Our visitor turned to me and asked if I was the Florek from Blogie.

'Why are you asking me this question?'

He said he had heard that in Blogie there was a partisan by the name of Florek.

'That's me,' I said.

Then he took the bottle of vodka, filled two glasses and asked me to drink. I asked why. He told me that he had been looking for me and wanted to kill me. I again asked why.

'I can't tell you. I don't know why myself. I can see that you're a good bloke and it would have been a big mistake to kill you.'

We drank our vodka and then he asked me if I knew Janek Skladowski. I said, 'Yes, I do. We drank a lot of vodka together. I met him at Stefania's house and at Mija's farm.' He cut short the conversation abruptly. If the stranger knew Janek, then I was sure he must be a partisan or from an underground group and posed no threat. He could see we were armed and understood who we were. However, we could take no risks and as soon as he had gone, we packed our things and moved off.

We got to Plebanka, a village near Blogie. I knew a friendly family farm where we could stay for a few days. From there, I went on my own to Blogie to visit Marian, but he was not home. Janina, as ever, was happy to see me. She asked how was I managing, how was my health; endless questions. I told her that I was now with a Russian group. I mentioned the stranger who had wanted to kill me. She said, 'You should be safe now. Janek was shot by the leadership of the AK. He was caught robbing farmers.'

I left Janina and went on to see Stefania. She, too, gave me a warm welcome. I asked her about the two men who had been after me and she said, 'Yes, they asked all about you. I told them that you were one of the partisans I got to know through Marian. You were friends who came for a drink and a chat. I knew nothing about anyone's private affairs.'

Fight to the Death

Back with the group in the evening, I heard how the Germans were targeting partisan groups. There had been a battle between a large AK group at Diabla Gora, a hillside forest where the Germans had followed the partisans for some days and surrounded them. The partisans in the hills inflicted serious losses on the Germans before escaping during the night.

German troops were facing one setback after another, especially on the Russian front. On 14 September 1944, we heard over the radio that the Red Army was approaching the River Wisla on the eastern side of Warsaw. The previous month, there had been an uprising in Warsaw led by AK fighters determined to liberate the capital before the Russians got there. They failed to dislodge the Germans.

We had to be doubly careful. Germans were now more active in the region, while the NSZ, the Polish nationalist group, had also become an increasing threat.

Mikhail, one of our Russian partisans, was drinking too much vodka. When drunk, he'd try and find a woman. I stopped him attempting to grab several. After one incident, he threatened to shoot me, shouting, 'You bloody Pole, I'm not taking orders from you.' Gromov disarmed him and warned if it happened again he would be shot. A report was sent to Kiev.

Taking refuge in a farmhouse near Alexandrow, our group settled down for the night, intending to move on the next day. I asked the farmer where we could buy a bottle of vodka and he told me that he made his own in his storeroom and gave us a bottle, refusing to take any money. Early in the morning, the lookout guard reported some movement of lorries and troops in the village. We prepared ourselves in case the Germans came towards the house; it was not safe to move into the forest because we were surrounded by open fields.

While we were hiding in the barn, Mikhail managed to slip away to the storeroom and put a glass out to catch vodka dripping from distillation pipes. When the farmer realised that Mikhail had been over doing it, he came to me and asked us to get him out of there. I told Major Gromov and he went and dragged him out and told him he'd already had several warnings and that he could be shot. Mikhail ripped open his jacket and shirt and shouted, 'Shoot. Shoot me then.' Gromov pulled the catch off his pistol and was ready to fire. I shouted to stop. The Germans were only 800 metres away and if they heard a shot they'd come straight away. The Major agreed and asked the boys to take Mikhail to the barn and keep him quiet. The Germans moved out of the village in the afternoon and we left after dark and moved towards Siucice.

We reached the ranger's house near the River Czarna, where we organized some food and most of the group stayed the night. Late that evening, Lieutenant Misha and I went to the village of Przylek to buy some meat for the group. We bought a small pig and after killing it, took it to a house where I had friends of my own age that I had known from before when staying in Siucice. There we cut up and prepared the meat to take with us back to the forest where we planned to meet the rest of the group and enjoy a lunchtime barbecue. In the evening, we were due to go to the ranger's house for dinner.

The preparation of the meat had taken all night. Lieutenant Misha and I were ready to leave just before the sun came up, but coming out of the house I saw some fires burning along the river bank at the edge of the forest. Alarmed, we looked through our binoculars and saw that the forest was surrounded by Germans. My friends' parents were frightened enough and I didn't want to put them at risk. We wrapped the meat in sheets and hid it in straw in the barn. Misha and I left and walked towards Siucice with our machine pistols hidden beneath our jackets.

Walking along the road, the field rose while the road dipped. Coming towards us from Siucice we saw two horses pulling an open carriage carrying four German officers. We decided that if they stopped us, we would open our jackets and fire at them. But they looked at us and carried on on their way.

Once in Siucice, we stayed throughout the day and in the evening went to the ranger's house where we reunited with our group. They had been in the forest during the day and Gromov, realising they were surrounded by Germans, decided not to open fire because it would be too difficult to escape over the river. The Germans could not see anyone through the bushes and trees and walked into the forest, shooting at random. With no return fire, they must have assumed we had already left, and withdrew. Someone must have alerted the Germans that we were in the area. I suspected it might be the ranger but had no proof.

Misha and I realised we had been lucky when passing the German officers. He said to me, 'Lewko, I'd rather it was me that got shot than you. You could survive without me, but the group would be lost without you. We'd share the fate of the first parachute group. You're the only one who knows how to navigate the forests and which farms to stay on. We'd have been shot by the NSZ by now.'

One evening, staying in a village, Mikhail managed to slip out and get himself drunk. He walked into the home of a disabled woman, intending to rape her. Someone in the house heard the noise and called her neighbours to help stop him. Two farmers came and told us what had happened. Major Gromov called a meeting of our five commanders straight away. Four of them voted to execute him. I was the only one not to put my hand up. One of the officers volunteered to carry it out. He went and found Mikhail in a house in the village, took him outside and shot him, and asked the farmers to bury the body.

Around that time, our group were staying at the farm where I had first met Major Gromov. The farmer and his wife were poor, but they helped as best they could. We repaid them with food and supplies. They had 11 children, the youngest a one-year-old boy. All the rest were girls. One evening, I went shopping with Lieutenant Misha by cart. I woke up the shopkeeper in Skorkowice and bought toothbrushes, toothpaste and other essentials. I visited Stefan's family in Chorzew and Siucice.

On the way back, I stopped at the mill in Rozynek for flour. I crossed the bridge over the River Czarna and went through the forest to the farm. It was early morning and I lay down between the others on the floor to catch some sleep, putting my head on Masha's comfy breast. She didn't mind; she looked on me as a brother and liked taking care of me. I fell asleep right away, but in a short while Masha woke me saying, 'Lewko, some partisans are coming.' Our guard had seen some men on a cart stopping on the track just where I had stopped months earlier.

Three men jumped off the cart and ran up to the neighbouring farm about 100 metres away. One stayed behind on the cart with the driver. I looked through my binoculars and saw that they were not partisans but Germans. I ordered my boys to climb out the windows that faced the forest, taking the children, the farmer and his wife. I was the last, and as I was about to peer out the open front door I saw a German running towards me. I shouted, 'Hands up', but he turned back and ran. I started shooting and hit him on the chin as he looked round. He fell over, stumbled back up and ran off into the neighbouring farm. I was out of ammunition and grabbed the rifle he had dropped and began shooting at the other two who were running towards the fields. I killed one and the other was shot by the boys who came from behind the farm. The injured German came out of the house, his head bandaged, and started to shoot at me with his pistol. I hadn't noticed him because I was shooting at the others and felt a bullet whizzing past my head. The boys gunned him down with their machine pistols.

The remaining German fled on the cart to the nearby village of Przylek, where a German Army unit was stationed. Within about 20 minutes, German lorries and tanks arrived, shooting into the forest but not venturing into it. We hid the family safely in the bushes and went a kilometre or two deep into the forest until we reached the River Czarna. We had to cross the river for safety. I was the first in the water, wading through to the other side. The rest of the group followed. The forest carried on and once away from the riverbank, we did our best to dry out our clothes. We sat down and ate our 'iron portions' – emergency bacon or dried meat that we carried in our rucksacks. One of the Russians managed to bring a bottle of vodka with him and we all shared a drink.

We caught sight of three farm workers taking a short cut through the forest and stopped them. They were on their way to the village of Rozynek for their lunch. I said they would have to stay with us until we left. We could not risk them letting anyone in the village know of our presence. Soon after, in the distance, I spotted the mill manager walking with a young woman. I went up to them and asked where they were going and what was happening in the village. I had met the manager the previous evening when he gave me some flour, which he refused to take money for. He said he was trying to avoid a Gestapo officer who had told farmers in the village to prepare their horses and carts. They were to be used in moving earth to build trenches. I told them both they would also have to stay with us until we left.

The woman said that she was on her way to teach children in the next village and they were waiting for her. The man said he guaranteed she would not tell anyone about us and asked if I would let her go, which I did. Then I asked the manager to go back to the village and find out more details about the Gestapo officer, where we might find him. It was unusual for a German to be on his own; he was obviously sure of himself. I said I would meet the manager at the edge of the forest in 10 or 15 minutes.

I took off my boots and poured out the river water. I could not get the boots on again because my socks were wet, so I put them on without socks. It was cold and I was angry.

Losing my family, being alone in the forest through that harsh winter, having men hunt and want to kill me had all taken its toll. I was depressed and couldn't see an end to the War, and even if the War did end, I didn't have anybody; I was frightened of the future.

Here, at last, was a chance to take my revenge on a Gestapo officer for the death of my brother. I didn't mind dying that way, in fact I wanted to die; at least I would die fighting.

I went back to the group and told Gromov that I wanted to get the German and bring him back with me. Gromov did not want me to go; he thought it was too dangerous and he was frightened that he might lose me. Then I asked the boys if any of them wanted to go with me. Wrobiov volunteered, and Gromov gave permission.

I armed myself with a machine pistol slung on a strap round my neck. To cover my pistol I took a leather coat from one of the farm workers, and his leather cap. Wrobiov did the same. We made our way to the edge of the forest. There, waiting for us, was the mill manager with a young man. They said that the German was alone. The mayor had given him some food and plenty of vodka. The officer had taken the farmers' identity cards so he would know if anyone did not show up. He told them to go and join the farmers from Ciechumin, the next village. The village was about 600 metres away and unusually long, with houses only on one side.

By the time Wrobiov and I reached Rozynek the officer had gone by cart to Ciechumin. I asked Wrobiov what he thought, should we go after him? He said, 'Definitely.'

When we arrived at the first houses in Ciechumin we saw farmers preparing their horses and carts. Women were crying. I asked one woman what was going on. She said a German officer was taking the men and horses to build trenches. I asked her where he was, she said with the village mayor.

We walked nearly all the way through the village to the mayor's home, purposely carrying on to the next house. I needed to find out what room the German would be in. The door to the next house was open and Wrobiov and I walked in. I saw men chatting in the kitchen. I acted green and asked where the mayor's house was. One of them said it was next door. 'But the mayor's here.'

I asked the mayor if he had a visitor. He said, 'Yes. He's a German officer and he is taking farmers to work with horses to dig some trenches.'

'Why are you here and not with the German?' I asked.

'I wanted a bottle of vodka for the German. He wants a drink.'

'So if we wanted some food, would you give it to us?'

'Yes, of course.' That was when it clicked and he realised who we were. He asked me to leave the German alone, not to do anything that would be bad for the village. I asked him what room the German was in and what he was doing.

'He's eating breakfast in the front room.'

'What did you give him to eat?' It wasn't that I needed to know; I just needed time to work out my move.

'My wife made him scrambled eggs with bread and butter and coffee, but he wanted some vodka.'

As we were talking the mayor's wife came in, calling him to come because the German was anxious. 'He wants you to bring the vodka.'

I said to Wrobiov, in Russian, 'Stay here and don't let anyone out.' I left the house and went next door. I opened the door and with my machine pistol in my hand shouted in German, 'Hands up.' He was sitting by the table in front of the window. He stood up and with his right hand reached for his pistol in his holster that was on his left side.

I pulled my trigger. One shot came out and missed, then the pistol jammed. With it still slung round my neck, I jumped on him and we struggled like two wrestlers. He was stronger, bigger and a head taller than me. I had the advantage of being driven by hate. I could not give in and allow him to wipe out the last Jew in the area. At that moment he had the physical advantage. He was pressing me down with all his weight. Through sheer strength of will I held his wrists. I was hindered by my pistol on my chest. 'The end must be near,' I thought. 'Help, Misha,' I shouted, confused.

I could smell alcohol on the German's gasping breath. I realised that he was losing stamina. When he managed to catch his breath, he pulled out his left hand from my grip and went for his pistol. I kicked his hand away. He stumbled on to his feet and as he did so, I used his weight to pull myself up. His holster was open and I grabbed his pistol. Once on my feet, I pushed him from behind on to the wall. I saw he was wobbly and unable to keep his balance.

I wanted to take him alive so that we could interrogate him, and pushed him out the front door. The fresh air immediately revived him. He tried to pull himself out of my grip and fell face down on the threshold, halfway down the steps, pulling me with him onto his back. I hit him on the back of his head with the barrel of the pistol. To my surprise the pistol fired; it already had a bullet loaded. By hitting him, I had released the safety catch. The bullet went through the back of his head. Hearing the shot, Wrobiov came and pulled me off. The German was still moving. Wrobiov finished him with another bullet.

With not having slept all night, the fighting, the running away, the crossing the river – together with a pain in my chest from my machine pistol pressed against me by the SS man and blood from my lips running into my mouth – I was exhausted and breathing heavily. I looked at the SS man lying on the ground with wild eyes and was glad it was not me; I had wanted to die but when it came to it, I fought for my life and now knew that I wanted to live.

A few farmers waiting for the SS man came along when they heard the shots. The mayor and his neighbour came in. I asked them to bring horses and a cart, load the body and cover it with straw. I told the mayor he could tell everyone they no longer had to go to work for the Germans. Wrobiov and I got on the cart and told our driver to head for Rozynek as quickly as possible. We passed a long queue of farmers with their carts waiting for orders from the German. They were surprised to get an order from me to go home.

On the way to the forest, I asked Wrobiov why he had not come and helped me earlier. He said that the mayor and the farmer had grabbed him after the first shot. They were sure I had been killed and said they did not want the German to kill him too. I was sure the mayor would rather have seen me dead, and if the German had killed me, they would have handed Wrobiov over to him. They let Wrobiov go when they heard the second shot; maybe the farmers told them that the German had been killed. They must have been disappointed to see me still alive.

We arrived at the forest and the boys did what they had to do with the body. We emptied the German's pockets and found he was an SS Oberleutnant; his documents were taken by Major Gromov. I gave the farmer all the farmers' documents so he could return them to their rightful owners.

Saving the Major and a German

We moved out of the forest avoiding villages, walking on the tracks dividing fields and carefully crossing main roads, stopping several kilometres on at the next forest for a rest before making our way back to Blogie after dark. We stayed over at Anton's farm, where we dried our clothes and ate. Next morning, we went to our forest base where we rested. We decided to split into two groups of eight each. Lieutenant Misha and myself were in charge of one group, and Gromov and Andrej the other.

Gromov's group went to stay at a farm near Siucice before proceeding to the forest, where they were due to pick up supplies dropped at midnight from a Russian plane. While at the farm, they were attacked by Germans.

Wrobiov, who was on guard outside, was killed. The group ran out and shot their way through the darkness. Major Gromov was wounded by three bullets in his belly and fell unconscious. The boys carried him to the nearest village, Ciechumin, where they took horses and a cart and brought him to Plebanka, where we had a contact and a place to meet. The Germans burned down the farm that had housed Gromov and his group.

One of Gromov's group came the few kilometres to Anton's brother's farm where we were staying and told us the news. I went immediately to Plebanka, where I found Gromov unconscious and in a bad way. Lieutenant Misha and I took a horse and cart from the vicarage in Blogie and went to a doctor I knew of in Dabrowa. The doctor's mother told me he was away and the nearest doctor was in Alexandrow, ten kilometres away.

In Alexandrow, the night guard directed us to the doctor's house. I knocked on the window and called out for someone to open the door, shouting that we needed help. Nobody opened up. I could hear whispering voices and decided to break in.

The doctor and his wife were in bed. He said he had not come to the door because he thought we were thieves; he had been robbed many times before. I described Gromov's injuries. He dressed and packed his emergency bag.

The sweat of our running horses glistened in the receding darkness as we arrived in Plebanka. The doctor took off Gromov's bandages and cleaned the wounds. Fortunately for Gromov, the bullets had skimmed through the side of his bulging belly and avoided his intestines. He was given an injection and medicine. The doctor told me we needed to regularly change the dressings. I asked how much we owed him but he did not want any money, asking only to be allowed to make his own way home; he felt safer walking alone. I warned him that his family would be punished if he said anything; he said we could trust him.

The next day, Gromov regained consciousness. He could not remember what had happened. I heard about another doctor I had known from Sulejow, Dr Gaida, who was now living in a nearby village. I told him I needed help; my wife was having a baby. He packed his bags and came with me. When we arrived, I gave him the dressings left by the other doctor for my injured friend. I took out my gun and told him that we were partisans. I said I would pay him for each visit, but it had to be kept secret, otherwise he would be risking his life. He agreed and refused to take any money and promised he would look after Gromov.

We needed to move Gromov for safety, in case the previous doctor told the Germans of our whereabouts. We carried him on a cart to a farm more convenient for the new doctor and left two guards.

It was November 1944 and winter was descending fast. The parachute drop should have supplied us with warm clothing, explosives, silencers for our guns and money. The plane did not make its drop because it did not get the signal from the ground (a triangle of fires lit in response to its flashing lights). We were unable to arrange another drop. It proved a serious setback, preventing us carrying out several planned actions.

We heard by radio that the Russians had instead dropped supplies to an AL group, who took over some of our actions.

We spent the winter of 1944/45 avoiding the Germans, unable to launch attacks. We moved from place to place only at night, staying nights in barns and days in the forest, catching what sleep we could. In the forest we would light a fire and lie around, only able to keep warm on one side, then turning to warm the other. On many occasions, as we were walking during the evening and night from village to village, I would fall asleep as I walked, stumbling into the person in front of me.

We never strayed too far from Major Gromov, who I checked on every day. After several weeks, he recovered fully and rejoined the group in early January 1945 as the Russian front moved closer.

One evening in Plebanka, one of our guards spotted a lone German soldier wandering into the village. We grabbed him and took him prisoner. His name was Herbert; he said he had come to the village for food and was not a Nazi. He claimed he had not killed anybody; his job was to drive horses. He was on the run with a group of deserters and did not want to fight. Hitler was 'Kaput', he said.

Major Gromov wanted to kill him, but I suggested that Karl and Werner should interrogate him. If his story stood up, we should let him live and stay with us. Gromov finally gave way.

It was obvious to me that Herbert was not a fighter or a Nazi, unlike the German prisoner that Lieutenant Misha and I had taken in a village near Blogie the previous summer. One Sunday, we were visiting a farmer's daughter I was friendly with, having a bite to eat and a few glasses of vodka. Her little sister came in and mentioned that a German was at another farm with girls from the village. I asked what German she meant. My friend said he came to the village some Sundays for butter and eggs and a drink with local girls. He was a sergeant in charge of slave labourers building trenches. I said to Misha, 'Shall we go and have a word with this German?' He agreed, and the girls showed us where to find him.

I opened the door of the farmhouse. He was sitting at the table with his back to us, surrounded by girls laughing and teaching him Polish. He turned around and shouted in German, 'Go away.' I opened my jacket, took out the pistol and told him to put up his hands. We took his gun and marched him off through the fields to the forest.

I reported to Major Gromov that I had brought a prisoner. He asked Karl to interrogate him. The prisoner said he was in charge of the Baudienst brigade digging trenches. He was married, had five children, and was a fervent believer in Hitler and National Socialism. Karl asked him if he knew where he was. He said, 'Yes, with bandits.' Karl hit him: 'You're the bandit, doing what you're doing to this country. You are a German, you should stay in Germany. I am a German and I am fighting fascists.' Karl asked him, 'If we let you go free will you carry on fighting against us?'

'Jawohl.'

Karl took out his knife, a souvenir of his fighting the fascists in Spain, and stabbed him through the heart. The soldier was tied to a tree. His head fell to one side as he took his last breath. Karl spat at him, 'It would be a pity to waste a bullet on you.'

Herbert was different. I believed his story and said we should let him live; everyone agreed. He wanted to kiss me and cried. He could not believe that his life had been saved. He told us that he had a wife and five children, he was a miner in Silesia and a left-winger. He did everything that we asked him to do, helped us keep camp and stayed with us in the village.

Two days after being taken prisoner, Herbert told me he knew where there were machine guns hidden in the forest and wanted to bring them to us. I told a sceptical Gromov, who was sure Herbert would run off. I didn't think so, but even if he did, he was no threat to us. Herbert knew his best chance of survival was to stay with us. Gromov relented and gave him his chance. A couple of hours later Herbert returned, loaded on both shoulders with machine guns. He was very proud of himself.

Germans started to appear in the forest, running away from the advancing Russian Army. We could hear heavy gunfire relatively close by. We felt vulnerable, as we were not strong enough to fight a large group of Germans if they appeared in the villages. Gromov said we should keep out of villages during the day and stay camouflaged in the forest.

One morning, we left a farm in Plebanka and went deep into the forest, where we rested around a fire. We left guards hiding with machine guns on the edge of the forest, where they spotted four Germans on horseback coming across the fields towards us. One of our guards ran to warn us and we put out our fire and took up our positions. When the Germans got close, ten of us surrounded them and shouted, 'Hands up.' They did as we commanded. We told them to get off their horses and to lay face down on the ground, where they were searched.

Under their coats, they were dressed in full uniform with their officer insignia and their pistols in holsters. We asked them to remove their coats and took their documents and pistols. Major Gromov told Karl to ask them in German who they were. As Karl was talking to them, Herbert told me in a quiet voice that they were SS men, which Karl soon confirmed.

Major Gromov told Karl to ask Herbert what should be done with such Germans. Herbert, who only spoke German, answered that they were murderers; he sliced his hand across his throat and said they should be, 'Kaput. It's people like them who are responsible for the War.' Karl told Major Gromov what Herbert had said. Gromov took his pistol and gave it to Herbert and told him to shoot them, which he did. Gromov then patted Herbert on the shoulder and said, 'Now you are one of us.'

Herbert was shaking. He knew that if he'd refused Gromov, he would be the one to be shot. He wanted to survive to see his wife and children. He hated what the Nazis had done and, like many German soldiers, loathed the Gestapo.

Gromov asked Masha to radio headquarters in Kiev to report on our action and find out the news from the front. We learned that the Russians would be reaching our area within days and should take great care until the Red Army could liberate us. It was a very dangerous period and, in the confusion, we could end up under Russian fire. We were told to make contact for further orders once the army had arrived.

The Russians Arrive

On 14 January 1945, Major Gromov said he wanted to meet up with the Russian Army as soon as possible. I argued that it was safer to stay in Plebanka, but he ordered us to move nearer the main road to Opoczno in the direction the Red Army was coming. That night we moved off, encountering groups of Germans who were sniping at us from the forest where they were now hiding.

We stopped overnight at a farm not far from the main road, where Major Gromov asked the radio operators to report our position to headquarters and find out the Russian Army's movements. They were told that Opoczno was free and the army was moving quickly towards Sulejow and the River Pilica. We were elated; waiting for the moment we would meet the Russians, looking forward to the end of our struggle.

At daylight on 15 January, we left the farm and headed for a village on the main road. We walked through a forest and from its perimeter could see the village about a kilometre away. We decided that Lieutenant Misha, Werner, who was carrying the machine gun, and I would carry on to the village, while Gromov and the rest of the group stayed in the forest, watching for a signal that we had made contact with the Red Army.

Halfway between the forest and the village, we suddenly heard artillery and machine gun fire coming from a forest on the other side of the village. Groups of German soldiers were running from the village straight towards us, heading for the forest we had just left. We lay down in the field and started shooting with our machine gun and machine pistols. Hearing the attack, the Germans changed direction and veered towards another part of the forest and disappeared.

The distant gunfire slackened as we continued on our way

to the village. All was quiet when we arrived and we carried on walking east on the main road. A Russian tank appeared from a bend and stopped right in front of us. A major jumped off the tank and came towards us with his pistol in his hand and, in Russian, asked us who we were. Misha told him we were Russian partisans and the major asked to see our documents. Misha took out from his watch pocket a piece of silk bearing the stamp of a Russian parachute group. (The silk kept the stamp intact even in water).

The major embraced us and went back to the tank to get some chocolate and told us, 'Your job is done. You should go and rest.' He looked at his watch and apologised for having to go, saying, 'Wait for the army, they'll be here soon.' He jumped on to the tank and waved goodbye as the tank moved forward.

At the end of the village, about a hundred metres away, the tank stopped and turned its gun towards the fields where the Germans had been running. The tank was taking up a pre-planned position and at the given time, gunfire started up again in the distant forest. German lorries, forced out of the forest, headed past the village and into the frozen fields, which they found difficult to negotiate. The tank blew up every one of them.

We went into the fields to look for any surviving Germans. We found one running towards us with his hands up and shouting in German, 'I'm not a fascist, I'm a Lutheran.' His left eye had been shot out and blood was pouring from the socket. I asked him if he had his bandage. He pointed to a little pocket inside the bottom of his jacket. I took it out and bandaged his head. We took him in to the village and told him to follow the other prisoners of war and he would be sorted out in Opoczno. He was grateful and thanked me.

Within minutes, a Russian jeep arrived in the village carrying a commander accompanied by three officers. An officer jumped out and asked who we were. Misha showed his identification. The commander said they were on their way to a manorial farm in the next village and we should join them. Misha told the commander that the rest of our group was in the forest; we needed to collect them, then we could join him at the farm. We told the commander that there were a lot of disarmed enemy soldiers hiding in the village. He said if they were Germans, we should send them on the road to Opoczno: if they were Wlasowa Ukrainians, we should shoot them.

We went into the houses where enemy troops were hiding and ordered them to march off to Opoczno. The Russians were ruthless with the Ukrainians, who they regarded as traitors, and we saw many bodies. It was a chaotic scene; I was aware that at any time I could get shot from a fleeing German who had not been disarmed, but I was too busy to be scared.

Outside the village, we saw the horrific sight of bodies crushed by tanks and lorries running over them. It looked like asphalt made out of flesh, blood and uniforms.

Columns of Russian soldiers poured into the village as the daylight faded. Misha, Werner and I at last had the chance to return to the forest to fetch our comrades, but could not find them. We carried on to the manorial farm, where we came across the tank commander. We were given food and drink and stayed overnight. We had arranged to meet Major Gromov and the rest of the group in Plebanka if any of us got lost.

At the manorial farm, the Russians were holding prisoner a group of about 30 disarmed Polish AK men. I recognised their commander, who I had recently met in the forest. I told the Russian commander that the Poles had been fighting the fascists and were sympathetic towards us. The commander said they could go home or join the Polish Army, though he would hold on to their weapons.

On the morning of 16 January, the day that Warsaw was liberated, the commander asked us where we were heading. We told him we wanted to go to Blogie. He took out his map and looked, then said he would give us a jeep and two soldiers to take us there.

In Blogie, Misha, Werner and I dropped in to see Janina and Stefania; Marian wasn't home. Stefania cooked us some scrambled eggs. Her mother arrived and said she had heard distant shooting and thought the Germans were coming back. I told her we had already met the Russians and they would be here soon. What I didn't realise was that there were still pockets of marauding Germans. We made our way to Plebanka, where we found Major Gromov and the rest of the group. We gave Gromov an account of our meeting with the Russian Army and he told us what had happened after we left them in the forest.

Seeing the Germans running towards them, they realised they were outnumbered and made their way to Plebanka and safety. While Gromov was talking, a guard reported that a German was coming our way on foot. I went out and as he approached shouted, 'Stop, hands up.' He made a run for it, but I began shooting and ran after him. He stopped and put his hands up with his back to me. I removed his pistol from under his coat and took him prisoner, putting him with other prisoners in the cellar of a farmhouse. After I locked him up, I realised there were no more bullets in my magazine. I started shaking with the shock of this realization.

As lunch was being cooked, our guard informed us that a column of German tanks was heading close. Major Gromov told us the same thing had happened the previous day. Everyone had stayed inside the house and the Germans had passed through into the forest.

I wasn't taking any chances. I went to the room opposite the kitchen and put some bullets into my pistol and machine pistol. The guard outside came in and stood behind the front door.

The tanks passed straight by. Then one stopped; a soldier jumped off and tried to get into the house. As he opened the door and stepped over the threshold our guard grabbed him, but the German managed to slip out of his hands. His fur hat fell off and he started to shout, 'Bandits. Bandits.' The boys started to run out from the back of the house. I heard the boom of a tank gun and was hit in the calf of my right leg and fell over. Fortunately, the leather on my long boot was so hard that the shell slid off, went through the old farmhouse's wooden wall and exploded in stones at the end of the yard. By some miracle, the tank shell only skimmed me.

The German machine guns started firing at our group. I heard one of them climb up the ladder in the hallway to the loft. I thought that would be my best chance too and was about to follow him when I heard one of them say, 'Pull the ladder up.' I was lying on the floor and wasn't sure if I could walk on my swelling leg. Instead, I pulled myself into a corner behind the oven and covered myself with clothes hanging on the wall. I covered my head with a sweater and could see out through the stitching.

I heard Germans coming along the hall and into the kitchen. They took away the two large saucepans, one full of potatoes and the other meat. When the shooting ceased, I could still hear German voices in the house. One of them came into my room. He looked around and went out. Another one came in, looked around, and also went out. Then a third did the same. When he left I felt relieved; I thought it must be three times lucky. A minute later, a fourth German came in. He looked around, saw the toes of my boots and ran out shouting, '*Noch einer ist da, noch einer ist da*.' [There's another one here]. I had no choice but to clamber up, break the window and climb out the back of the house. At the rear exit to the yard I saw Werner and the wife of the farmer lying dead; they must have been killed by the exploding tank shell that had bounced off me. I ran, limping, along the frozen field towards the next village.

The Germans started shooting at me from the tanks with their machine guns. After about 200 metres I fell on the ground. I couldn't carry on. My right leg was swollen from the shell and my left knee was dislocated.

When I tried to look back I had to shut my eyes because the frozen earth thrown up by the bullets was slapping my face. When the guns stopped firing, I saw a German at the back of the village running at me with his rifle. I didn't know whether to turn round and start shooting at him or carry on running. When I tried to move, I realised I couldn't walk. I had to shuffle along on my elbows and knees. The machine guns started up again and the German moved back to avoid the bullets. He must have believed I was done for. I could hear shelling; it sounded like the Russians were coming. At this, the Germans made off.

A little way ahead of me I saw a barn had been set on fire by bullets aimed at me. Farmers were trying to put out the flames, while others were escaping on horses and carts. They heard me call out and one of them jumped off the cart, picked me up and drove me away with them to the next village.

In the evening, our group reunited and we tended to our

seven wounded men. I could only walk with a stick. We stayed in the village and put a guard at its eastern entrance. During the night, someone appeared directly in front of him. It was so dark that he could not make out who they were. The guard pointed his machine pistol and shouted, 'Stop. Who's there?' The man did not reply and grabbed the machine pistol by the barrel. The guard pulled his trigger, shot and killed the man. When he examined the body he found it was a German soldier. Hearing the shot, the boys ran out and moved the body out of sight of others who might be coming up from behind.

I realised it would be safer if we moved to the middle of the village; we'd get an earlier warning if they came in from either end. Gromov agreed. We carried the wounded with us and found a cellar to keep them safe. We took bedding from several houses and made the wounded comfortable, telling them we would come back the next day. On our way through the village we had come across five more Germans, wandering on their own and looking for food. We killed them all.

The remaining eight of us headed for Sulejow to make contact with the Russian Army and get help for the wounded. We walked four or five kilometres to the next village, where we ate and rested for a few hours until it was light. From there, it was only three kilometres to Sulejow.

We walked along a track by the river. Five hundred metres out of the village we heard machine-gun fire on the other side of the frozen river. A large number of German soldiers were fleeing the fire, running on the ice towards us.

I suggested to Major Gromov that we should all take cover. He disagreed and commanded that we all lay down immediately and open fire. We shot at the Germans as they were running over the ice. When they heard our machine guns they began to run down the river and disappeared.

We realised it was too risky to go on the open road to Sulejow. Instead, we went up a hill and started to walk through some trenches. There we met a group of Germans coming the opposite way. They did not fight and put up their arms to surrender. We ordered them to leave all their weapons and to walk along the river to Sulejow, back from where they had come. On the road, Russian soldiers took them to join the other PoWs.

We abandoned our journey, deciding it was better to go back and take care of our wounded; but, when we arrived back in the evening, it was to find that they had already been taken to hospital by the Russians. We carried on to Blogie, where we stayed on Anton's farm. The village was occupied by the Russian Army. Stefania's flat, next door to the police station, had been taken over by the Russian commander. I went with Gromov to introduce ourselves as a parachute group. They treated us well, giving us vodka and food.

The next day, I was asked to go with four officers by car to Sulejow. On the way, we saw a group of Germans running from the forest, crossing the road and a field to reach some woods. The major in our car ordered the driver to stop, and we ran after the Germans to take them prisoner. We told them to stop, but they did not listen.

Some of them managed to reach the woods, but three of them stopped. Two of them dropped their rifles and put their hands up. The middle one held on to his rifle, though it was pointed down. The major wanted to shoot him but his pistol did not fire. The German fired his rifle and killed the major. At this, we ran to surround them. The Russian officers tried to fire their pistols but they did not work. I started to shoot from my machine pistol and killed all three Germans. We carried the major's body into the car and drove back to Blogie, where he was buried.

The commander thanked me for my action, and for saving the lives of the other officers. They asked me to join them and go with them to Berlin. I wanted to go with them, but Major Gromov said, 'No. We still have a job to do.' He had received orders to take our group to Cracow and report to the Russian partisan commander. We were to prepare for a parachute drop into Austria. I was still limping and in great pain and asked the Russian Army doctor to fix my dislocated left knee, but she could not manage it. I went to Janina and she asked a healer from the village, who was also no use.

I started to organize our 500-kilometre journey to Cracow. We had four horses that we'd taken from the Gestapo officers, and two carts with two pairs of horses taken from the German Army. We packed supplies of food and I said goodbye to my friends in Blogie.

Major Gromov, the two lieutenants (Misha and Andrej), and myself were riding the horses. I had to be helped onto the horse. The other 11 boys and Herbert went on the carts. On the way to Cracow, we came across groups of Germans firing at us from the woods. We fired back with our machine guns mounted on the carts, taking some prisoners, marching them in front of the carts and delivering them to the Russian Army at the next town.

It struck me how the Germans had taken our place in the forest. It was they who were now hiding from us as we took to the roads.

Gromov was a Cossack and he showed off his equestrian skills, sliding off his saddle and going under the horse while it was trotting along. We took the horses cross-country so that they did not have to go on the hard road all the time. As I jumped over a trench I heard a sharp crack. My knee had locked back into its socket. I was shocked, but the relief was enormous. I could move my leg and the pain started to subside.

The journey took us over a week; we slept in villages and were in high spirits. About five kilometres outside Cracow we stopped at a manorial farm. The owners had fled and all the cattle had been taken by the labourers. The only people left were workers who lived in the farm cottages. Gromov and I told the rest of our boys to stay at the farm while we went to Cracow to receive our orders.

At headquarters, they asked us to take rooms in a block of flats and wait for further orders. For our food, we would have to go to the army kitchen. I suggested Gromov ask the commander if we could stay at the farm, as it was more comfortable there and they would not need to feed us; we could keep in touch by radio. It was only half an hour away. The commander agreed.

I organized the farm and ordered the workers who had taken the cows to bring them back so that we had milk. We sold the four SS horses in a market in Cracow and used the cash for supplies. We stayed at the farm for a month, until the middle of March 1945. It felt like a holiday, not having to worry about fighting. We were told that we were not going to be parachuted into Austria after all. The front had moved much faster than expected. The Russians in our group were transported home and Herbert was taken to a Prisoner of War camp, where he was given the privileged job of helping the camp commander.

I was angry with Gromov for refusing me permission to join the Russian Army assault on Berlin. I had hoped that from there, I would be able to stay in the West once the War was over, maybe get to America. Instead, I had to report to the Polish Army. They gave me papers proving I had fought with the Russian partisans; I was allowed free travel throughout Poland. There was no chance of leaving Poland now. First, I decided to go back to my hometown to see if any of my family or friends had survived.

Liberation

My war ended on 20 March 1945, although the official end didn't come until 8 May. I was supposed to join the Polish Army, but before I did that, I decided to go to Lodz. I knew that Sulejow had been destroyed; there was nothing to go back for. I had family in Lodz, where there had been a large Jewish population. I expected to find the remnants of the community.

I couldn't afford to stay in a hotel, but I had Stefania's address and went to visit her and her family. They greeted me warmly and offered to put me up. Stefania's father had been a policeman before the War. When the Germans came, the police were supposed to carry on with their duties, but he refused and left Lodz with his family to settle in Blogie. He was discovered and arrested and sent to a concentration camp and never returned home. Stefania's brother had been killed fighting the Germans in the mountains with an AK group. The family had been very kind to me during my visits to Blogie as a partisan, and Stefania knew about my Jewish origins, although we never discussed the matter.

Now that I had a base, I set off to the centre of Lodz to begin my search. I saw a man who looked Jewish walking in the street. I stopped him and asked him if he knew of a Jewish committee. I told him that I was Jewish and looking for my family. He gave me an address where I could go to register myself as a Jewish war survivor and get help in tracking my family.

I went and registered, but could find no trace of my family in the records. I went in search of anyone who might have survived, like the Rosenbergs, my aunt and uncle and their five children. I went to visit their home that was part of the ghetto in the War. A Polish family now lived there.

At the records office, I did find the name of a neighbour from Sulejow, Aaron Yurkiewitz, who was renting a flat in Wolborska Street in Lodz. I called on Aaron; we were overjoyed to see each other again. We cried over the families we had lost. He asked me where I was staying, although he didn't invite me to stay with him.

I settled down at Stefania's. As there was only her, her mother and sister, they were pleased to have a man in the house. At the weekends I would go to visit Aaron and gradually got to know his circle of friends.

At the end of May, I presented myself at the headquarters of the Polish Army in Lodz as a former partisan. I handed over the documents I had been given at the Soviet partisan headquarters in Cracow. I was registered and told to report to the army in Gorzow at the end of June.

I decided to visit my friends in Blogie and recover the photographs I had hidden in a barn. I visited several friends who had been connected with the underground. At Janina's home I met Marian Kubiak. I knew him from before the War. During the War, when I was in the AK, he was in the NSZ. He was now in hiding because he refused to surrender to the new communist government.

It felt like the whole village wanted to be my friend and drown me in vodka. They still did not know I was Jewish and I saw no point in telling them.

On my return to Lodz I went to visit Aaron, who I had not seen for a few weeks. When I rang the bell a strange woman answered the door. I asked if my friend was home. The woman replied that she was living there alone; she didn't know anything about Aaron. So I went to see one of Aaron's friends, who told me that he had left Poland and had gone to Munich, where Jews were assembling before being transported to America and Palestine. He had been looking for me and had wanted me to accompany him. He had read in the newspapers a story about two young Polish Jews who had survived the War and returned to their old homes in small towns. They found that Polish families had taken them over. The families had got them drunk and then, while they slept, had murdered them. Aaron was convinced that I had been one of those victims after he had been to see Stefania, who told him I had gone to the village and she didn't know when I would be back.

I still had some hope that one of my relatives in Lodz, Sulejow or Skierniewice had survived. With free travel, I could go to these places and look for survivors. I found no one, and as I was required to join the army I gave up thoughts of leaving Poland. I hoped there would be a better life where Jews were tolerated. I believed the communists when they said there would be democracy and freedom for all.

Stefania informed me she was pregnant, and I thought the best thing was for us to get married and take care of the child. Kamila was born on 18 December 1945.

I reported to the Polish Interior Army in Gorzow, a town that had historically been in Poland, then became part of Germany for many years and, now the War was over, was Polish once more.

I was conscripted for my two years national service. By now I was 23: conscripts are normally 21. I was given the rank of sergeant, owing to my partisan record, and awarded the Partisan Cross and one of the army's highest honours, the Military Cross. I was disappointed not to have been made a lieutenant with the rank of officer. I thought maybe it had something to do with the fact that I was Jewish, but I didn't want to admit that might be the case in the new Poland. The excuse they gave was that I had not been given the proper army training and at the end of 1945, I was sent off to Army College.

Despite the German surrender, peace did not descend over Poland overnight. There were still bandits operating out of forests and villages. They would rob people, attack police stations and army units. The bandits included groups of German soldiers who had refused to surrender. They hoped to make their way back into Germany. There were also Polish members of the NSZ who did not recognize the Polish Government. They targeted and killed local communist leaders. The most desperate bandits of all were former members of Wlasowa's Ukrainian Army, who had fought alongside the Nazis. They could not return to the Ukraine, where the Russians would have killed them, so they tried to survive in the forest, hoping the political situation would change.

The army made use of my experience as a partisan and gave me a unit of 25 men to command. Our first action was near Kostrzyn, close to the German border. A former nationalist sympathizer was operating as a bandit. We heard through intelligence that he was expected to visit his family, and searched his home but found nothing. I decided to wait in the house with some men in case he turned up. During the night it turned cold and I told the men to light the stove. As the wood started burning we heard shouting, 'Put the fire out, I'm coming down.' Our bandit had managed to hide away in the chimney.

By now it was the latter half of 1946, and it felt the most natural thing to be fighting again; it was as if my war had never ended. Over the next year I was involved in several actions, not so dissimilar to what I had been engaged in as a partisan, though during this period I did not shoot anybody and none of my group were injured. We arrested a few Germans and Ukrainians, but mostly they were individual Poles who we handed over to the police.

In 1947, a group of Ukrainians ambushed and murdered General Swierczewski, head of the Interior Army. It provoked the government into joining forces with the Russian Army to eradicate remaining dissident factions. Whole villages, believed to be sympathetic to the bandits, were emptied and their inhabitants moved to western Poland. By the end of 1947, all terrorist opposition to the government was successfully wiped out.

I was transferred to Poznan and given the job of quartermaster in charge of food supplies, staying there until 1951. On 2 December 1951 my son, Zbyszek, was born in Lodz where Stefania was staying with her mother while I was travelling on behalf of the army. In that period, the army sent me to the University of Warsaw where I learnt economics, politics, psychology, German and Russian.

I was offered a job in Slubice, on the border with East Germany, because of my knowledge of German and Russian. I was commander of the border and had risen to the rank of captain. I worked as a translator between the armies on each side of the border.

I bought a villa in Slubice, where I settled down with the family and was active in the local council and became president of the regional Combatants Organization. In order to serve on these bodies I had to be a member of the Communist Party, but I was never a communist. I was a democratic socialist. I expressed views that were often opposed to the party line that made some people uncomfortable.

I was keen to assimilate into secular Polish society. After the War, I had reverted to my original name of Lajbcygier, but when I took the post on the border, I was asked by army personnel to become Mayevski again, so that the Germans would not see my Jewish name. They preferred that the Germans did not know I was Jewish in case they felt the Poles were trying to humiliate them. I agreed for several reasons. I was aware of growing anti-Semitism in the army and Polish society and I did not want the children to have any problems at school. It became clear to me that my involvement with the AK partisans had left a question mark in the mind of the army over my loyalty, as had my Jewishness. I felt this was holding me back from being promoted to higher ranks.

I retired from the army in 1964 on a full pension, which gave me a comfortable standard of living. I was still involved as an elected representative in local government, was made speaker in the local assembly and was president of the Combatants Association.

Even after retirement, I was obliged to do four weeks annual service in the army as an officer. In our free time, fellow officers told stories of their experiences. Once, while I was billeted in Kielce, a local officer, not knowing that I was Jewish, told the story of the notorious Kielce pogrom of 42 Jews in 1946. A story had circulated in Kielce that Jews had kidnapped a Catholic boy, murdered him, and used his blood to make matzos (unleavened bread). Kielce's surviving Jews all lived in one block of flats overlooking the small River Silnica. A lynch mob, armed with sticks, organized a march to their building. When the Jews saw them coming and shouting, 'Kill the Jews,' they blocked the entrance, but the mob marched in. They broke into every flat and dragged out the Jews, men, women and children, beating them to death and throwing some into the river from the upper floor windows. It was a river of blood.

During the pogrom, a young Jew went to the police station shouting, 'Help, they're killing the Jews.' The policeman on duty outside took his rifle and shot the Jew. Later, he was sentenced for the killing along with some of the other murderers. The missing boy came home the next day. He had been staying with his grandparents.

To the officers it was just a story; they expressed no feeling on whether it was unjust. Some of them had witnessed it but did nothing.

A New Wave of Anti-Semitism

The Six Day War in 1967 and the Israeli victory over the Arabs changed everything, sending a wave of anti-Semitism, fanned by the Communist Party, throughout Poland. Local parties were instructed to check on Jews in their area to see if they were working as a fifth column against the state. My local party was informed by the regional party that I was a Jew. They were surprised, and notified all local executives of the fact.

A meeting of the Combatants Association was called, without my being notified, by my deputy (who was also secretary of the local party), to discuss him taking over my presidency. At the meeting the deputy said, 'We don't need a Jew for president. Don't we have enough Poles?' One of the veterans stood up and said, 'What's wrong with Mayevski? So what if he is a Jew, he has helped more people than all of us.' I reported the situation to the regional president, who asked me to ignore it. He didn't think they were being anti-Semitic.

I was called in to party headquarters in Slubice and asked to write an article for their newspaper, denouncing Israeli aggression as a Jew and former partisan. They reminded me that Poles had helped me survive. I agreed that they had, but I also pointed out that I had helped Poles survive and had fought for Poland. I told them about the Polish traitor who collaborated with the Gestapo and his list of fifty-six names of farmers who had given money to support the partisans. I refused to write the article.

As I was walking in the street in Slubice I met a good friend, a colonel who had worked in the army with me on the border. We were very close, so when I met him, I gave him a friendly greeting. He looked around and said, 'Sorry, Florian. I have to meet someone, I've got to go.' I realised that he was warned about befriending me and was frightened in case someone saw him talk to me. I was shocked and said to myself, 'I don't want anything more to do with this country. I have to find a way out.'

Another friend, named Smajewski, who I was very close to and called my brother-in-law because our names were so similar, was ordered to report to a police station. There they asked him what relationship he had to me. He said we were friends. He had to return to his place of birth, a few hundred kilometres away, and bring his birth certificate from the church to prove he was not Jewish. His wife told me about it. He did not want to say anything in case it embarrassed me. He told me that whatever happened, I should remember I would always be a friend of his family.

The following year, I wanted to visit a Polish friend in England. He was a Pole who I'd recently met and become friendly with, who had gone to England with the Polish Army from Russia during the War. In the back of my mind, I thought about checking out the possibilities of emigration once outside Poland. He sent the necessary invitation and ticket but my request for a passport was refused. There was some rule about needing five years after retirement before getting a passport to go to the West, but I suspected that the reason was because they did not trust me. They were probably right.

Instead, I went for a six-week holiday to Russia in response to invitations from Major Gromov, Misha the mine-layer, and a Russian combatants organization. It was a visit that gave me a real insight into the communist system, and I didn't like what I saw. The political repression was far worse than in Poland and the standard of living much lower. I had to get a special permit in order to visit the Black Sea, where I stayed at a hotel. I had to hand in my documents on arrival and they were handed back to me when I left. Papers had to be checked every step of the way.

From there I went to visit Gromov, who lived in the Tambov region deep in southern central European Russia. I arrived with Stefania at the train station in Rostov on my way to Gromov. From there I had to buy a ticket to Tambov, about 800 kilometres north.

It was so crowded that they warned me it would take days

before I could get to buy my ticket. I asked a policeman to help and showed my papers from Moscow from the Combatants Association headquarters. He got the stationmaster to get me my ticket. I asked the policeman to send a telegram to Gromov with details of my arrival at Tambov station the following day.

Stefania and I arrived at Tambov to find no one there to greet us. I left her at the station and went to look for a police station. I saw two policemen dragging away a drunk and followed them. At the station, I told them of my predicament. The station officer took me into his office and told a colleague to call up the farm office where Gromov worked to arrange for us to be picked up at his local station.

A newspaper reporter, alerted by the station officer that a partisan was in town, turned up to interview me. I was surprised I was big news and told her I couldn't talk, as I had to get back to my wife at the train station. She said it would not take long and I told her my story for half an hour. The article appeared the next day on the front page.

The police helped get my new tickets and took me and Stefania to another train station to catch a local line to a town near Gromov's village, where he picked me up. It was an emotional greeting. He was crying, and hugged me and thanked me for coming to visit him. He never thought he would see me again. Gromov's 90-year-old mother came to his house to meet me. She knelt down, held and kissed my legs, crying, 'You saved the life of my son.' She could not stop saying thank you.

Gromov was a veterinary doctor for state farms. He lived in a tiny village. In honour of my arrival, he paved the dirt track from his house to his perimeter fence and painted it red. And painted the wooden floor in his living room red. It was his way of laying out a red carpet for me.

That first night, he opened a wardrobe in the living room and took out a bottle of vodka. I saw that the bottom of the wardrobe was full of bottles of Moskevskaya vodka. I asked him if he had a vodka business. He said, 'No, there are only 40 bottles.' In the pantry, his wife had another 40 litres of home-made vodka. I was staying a week and all the vodka was there especially for my visit and to entertain the party heads and farm bosses who had been invited after a meeting where I was due to speak.

Despite his experiences in the War and his profession, Gromov was very unworldly. He was recently given a television, but it had never been turned on. I asked him if he would put it on and he said, 'Help yourself.' He simply had no idea of how to turn it on. When I turned the switch and selected the stations, he stood right behind me with his wife to see how it was done.

The local communist party asked the head of the local Sovchoz collective farms to organize a meeting so that Gromov and I could relate our war experiences. They brought in about 400 people from four villages. A regional communist official explained how Gromov had fought behind the front in Poland and how I had saved his life. He asked Gromov to speak first, but he was too choked and asked me to start. I excused my poor Russian and asked for a translator, but the crowd urged, 'Speak, speak, we'll understand.' I spoke for an hour. The Germans had never got as far as their region and their knowledge of what had gone on was limited. They liked to hear stories of Russian bravery.

After the meeting, Gromov entertained party and Sovchoz bosses at his home. The table was full of sausages, chicken and other meats brought as presents by the farm workers. We ate and drank until the early hours and discussed our life in the Russian partisans. To them, I was a hero, saving the lives of Major Gromov and the five Russian prisoners of war I took into our group. We ended up comparing farming methods in Russia and Poland. I pointed out how inefficient I thought they were. They respected what I had to say.

Then I went to visit Misha in a village near Kiev. Misha was a mathematics teacher at a college. I hadn't seen him since I put him in the cellar after he was wounded. It was another emotional reunion. I stayed with his family and he organized for me to talk to Communist Party activists and at schools, where they gave me gifts of books.

On one day, Misha took me on his motorbike into town. He left the bike with a woman friend for safety. As we walked away from her he turned and said, 'She's a good woman, even though she's Jewish.' He never knew I was Jewish and I did not say anything.

On my last evening, the party organized a meeting in a restaurant, where I gave a talk. There was plenty to drink: vodka, cognac and Russian champagne. I was giddy with the mixture of drinks and had a train to catch, taking me back to Poland. The party arranged for a driver to take us to the station. He took a short cut down a track to a flimsy wooden footbridge. He made Stefania and me get out and walk over, and then slowly drove the car over the weak bridge that shook under its weight. On the train, everything was spinning before me as I lay down.

After I returned to Poland, the impact of the anti-Semitism there became worse. I was forced to resign from my public activities, former friends began to shun me, and I even got anonymous letters telling me my real place was in Palestine. I decided that the time had come to leave Poland and make a new life in the West. The problem I faced was that I didn't know anyone in the West who was prepared to give me an invitation. A friendly local hotel manager suggested I could earn extra money by letting my spare rooms and empty garage to visiting foreigners. As a border town, we had many visitors from all over Europe come to stay overnight. In the summer of 1969, I got a phone call late at night from the hotel, asking if I had spare rooms for three people. I told my manager friend to send the visitors over. A couple and their fifteen-year-old daughter arrived from Belgium, their car fully loaded with luggage. I prepared some drinks and snacks for them. The father, Wladek, was Polish, and over a bottle of my home made wine I asked him where he was going.

'Siucice,' he replied.

'Who do you know there?' I asked.

'My mother.'

'What's her name?

'Laszczykowa.'

'Is that the widow who lives with her sister in the farm outside the village?' I asked.

He was astounded. 'Yes,' he said. 'Do you know her?"

'Yes,' I replied. 'I lived there during the War. When you visit your mother, please give her regards from Florek. She'll know who I am.'

Wladek was on his first visit to Poland since 1939. A new recruit in the Polish Army, he had been captured and taken to Germany as a PoW. After being liberated by the Allies he was taken to England, where he rejoined the Polish Army and took part in the liberation of Belgium. There he met and married a Belgian woman and settled down. Before the family left the next morning, he asked if he could book the garage and a couple of rooms for a month's time, on his return home.

When they next arrived, I went out to greet them and opened the garage doors. Wladek came rushing up to me, embraced me warmly and kissing me said, 'Now I know everything about you.'

In the village everyone had told him, 'He's the best. There is no boy like Florek.' They knew I was Jewish.

'My mother told me about the cat stealing your bacon.' I laughed and remembered how one evening, with a group of 12 partisans, I took out a piece of bacon from my rucksack and asked his mother to make scrambled eggs to go with it. She took the meat, put it on the table and went to fetch eggs. But the cat jumped up on the table and ran off with the bacon. I had to share another partisan's portion.

We drank vodka together until late. I asked him if he would send me an invitation to visit Belgium. I suggested that, as he worked at the Ford car plant, he should write stating that he had a car that he would like to give me. He readily agreed, and about a month later I received his invitation.

After I applied for a passport, I was called in for an interview by Sluzba Bezpieczenstwa (the Polish security service), as I was an ex-officer in the army. They asked, 'How long do you need to go?' I said, 'Oh, for a week. That'll be enough to go and collect the car and come back.' They wanted to know about Wladek, what he did and where he lived. They told me to be careful. 'You are Jewish, and Jews will try to find out things from you. You should not speak to anybody.' I told them I was going to a small town and there were no Jews there and that I didn't want to get into political discussions. My passport was granted.

I became estranged from Stefania, although we still lived together. At her work, she had people saying to her, 'Couldn't you find a Pole to marry instead of a Jew?' I didn't see why she had to suffer for me. Stefania knew I was leaving but we never discussed the matter, even though she knew I would never come back. Kamila and Zbyszek were grown-up and I hoped that, one day, I would see them again. I did not tell them I was leaving for good, though they had a good idea.

Becoming a Refugee

On the train through East Germany, I was still not sure I would get out. All it would take was a phone call from Stefania to the authorities and I would be arrested. It wasn't in her nature to inform on me, but she could have let it slip to somebody. As the train drew into a station in West Germany, I saw a Coca-Cola machine. I sat back and sighed with relief. 'Ah, now I'm free.' I knew I had made it.

I arrived in Brussels in September 1969, frightened of what the future would hold. Would I be able to work? I did not know which country I wanted to settle in. I first went to stay with friends in Brussels. There I decided immediately to apply for political asylum. I was granted a United Nations passport. After a month, I went to see Wladek in Stekene near the Dutch border, and stayed with him for several weeks. He was very hospitable. I told him I had been granted asylum and would not return to Poland. I was concerned not to make any trouble for him and his family. I explained my situation and that I had nothing to go back for; how upset I was at having fought for Poland in the War, and now I felt like I was being pushed back into the forest. He understood and said, 'Do what is best for you. Don't worry about me. Even if I have some trouble, I'll manage to get out of it.'

I returned to Brussels and stayed with my friend, Roman Dawidowicz, and his wife, Jadzia. Roman and I went to school together in Sulejow. During the War, he and his brother, sister and brother-in-law went as volunteer workers to Germany. The Nazis gave preferential treatment to Poles who volunteered to work in Germany. Their living conditions were better and they were relatively free, living in farms or with German families. Roman and his family did not look Jewish and took the names of Polish friends who had registered as volunteers for work in Germany. Posing as their friends, Roman and his family used the documents to become volunteers. As the documents had no photographs it was easy for them to pretend to be someone else.

One of the Poles that Roman worked with in Germany realised there was something strange about him. Like many Jews who spoke Yiddish at home, Roman spoke Polish with a slight accent. The Pole wanted to inform the Germans, but was told by the rest of the group that if he said a word, he would be dead.

After the War, I was on a tram in Wroclaw and saw Roman's brother, Janek, walking in the street. I jumped off and tapped him on the shoulder. He turned and said, 'Moshe Aaron. Is it you?'

He took me home and we went to visit Roman, who was living nearby. Some years later, Roman was able to leave Poland to join another brother who had emigrated to Belgium before the War. After Roman emigrated, he would visit me every year on his way to see his wife's family. I asked him if he would send me an invitation so I could leave Poland, but he refused, saying he was worried that he and his wife would not be able to come and visit the family.

Roman and Jadzia put me up until December 1969, when I went to London to visit Janek Kowalski. I returned to Brussels in January 1970. My friends helped me get my own apartment and I spent two years working as an interior decorator and carpenter and working for Roman in his furniture business. I didn't know where I wanted to settle. Before I made a final decision, I decided to go to Israel at the end of 1971. I thought I might find some surviving relatives. I had friends in Netanya and Bat Yam, and stayed in an ulpan (college) in Jerusalem to learn Hebrew. I visited the Holocaust memorial at Yad Vashem and tried to trace some relatives, but without success.

In Netanya, I stayed with a friend I met while visiting Roman in Wroclaw. Najmark had owned a small plastics factory, built a house, and had a good life with his wife and two children in Poland. He had been forced to leave Poland in 1968 because of the anti-Semitism. Everyday while staying with him I would go down to the beach, swimming and sunbathing near the Ein Hayam Hotel. I did not know it belonged to my cousin, Max Rosenberg, and was not aware he had survived the War.

In Bat Yam I stayed with Fiszel Szpiro, a neighbour from Sulejow. I was chatting to him one day in his shoe repair shop when a friend of his came in. I was introduced as a friend who had been in the Polish Army after the War. The friend told me he had a friend in the Polish Army who had been killed in a car accident in 1957, before he was due to leave with his family for Israel. I asked him if his name was Janek Shiller. 'Yes,' he replied. 'His wife, Tola, took the two little girls and had come to live in Netanya.' He gave me her address.

The next afternoon, I arrived at Tola's home in Netanya to find only her daughter home. I introduced myself and she said her mother would soon be back from the hospital, where she worked as a nurse. Tola arrived home to be told she had a visitor. I came out of the living room into the hall. Tola became excited and shouted in Polish, 'Don't tell me, don't tell me.' Then my name came to her. 'Florek.'

'Yes,' I said. 'It's me.' We cuddled, cried and laughed, remembering her husband, Janek.

I had been a friend of Janek Shiller in the army ever since he had driven me on an official trip to Lodz. We had got on well, and he invited me for a meal at his home one Friday evening after our return to Poznan. His wife, Tola, served up a traditional Jewish meal. I told them that I was Jewish, too. They were amazed, and it transpired that during our trip to Lodz, both of us had gone off separately to visit Jewish friends, both concerned that the other shouldn't discover this.

After six months in Israel I returned to Brussels. Two weeks later I received a letter from Aaron Yurkiewitz, my old neighbour who I last saw in Lodz in 1945, and who had emigrated to America.

It said: 'I am writing to you on the plane returning to America from Israel. When I arrive in New York I will post this letter to you. I wanted to know if you are the same Moshe Aaron who was at my flat in Wolborska Street in Lodz in 1945? If you are that same person, let me know and I will give you the address of your cousin, Max, from Lodz. He survived the War.'

The night before getting on the plane back to America, Aaron and his wife had been staying in Bat Yam with Fiszel

Szpiro, who he had not seen since before the War.

As they were chatting over some food and drink about who had survived and perished from Sulejow, Fiszel mentioned to Aaron, 'You know who was here? The widow's son, Moshe Aaron.'

Aaron said, 'I know he survived the War with the partisans. But he got killed before I left Poland. I read in the paper in Lodz that two Jewish boys were killed by Poles. When he never visited me like he said he would, I thought he was one of them.'

Fiszel said, 'You're wrong. He was sitting here at this table a few weeks ago. I'll get you his address in Brussels.'

Aaron's letter had a tremendous impact on me. Suddenly, I realised that I wasn't alone in the world. Another member of my family had survived, too. I was elated.

I replied immediately, confirming it was me, and asked for my cousin, Max Rosenberg's, address. I was surprised to find I knew it. The Ein Hayam Hotel, Netanya

I wrote to Max, who responded straight away with the news that, the following Thursday, he was flying to Vienna, where he had a wholesale clothes business and an apartment. His married daughter lived there and owned her own clothes shops. He wanted to meet me there. I could not get a flight and phoned him. He said he would meet me at the train station. I arrived to find no sign of him; somehow, we managed to miss each other. I got a taxi to his home; no one was in. Within minutes, he drove up in his Mercedes limousine. There he was, the only other survivor from my entire family. It was our first meeting for 33 years.

Our feelings were indescribable as we laughed and cried and reflected on how we were the only two left from a family of 40. We talked late into the night, relating our stories of survival. Max had been in Auschwitz, working as a forced labourer in a coal mine. His older brother had died in his arms in the mine. The rest of his family, two younger brothers, his sister and parents, had all disappeared after the liquidation of the Lodz Ghetto.

Max suggested I came to Israel straight away to live with his family. He said he had enough money; I didn't have to worry about anything. I thanked him, but said that I could not go straight away; I had to sort out my affairs in Brussels. Two weeks later I was in Israel, and once again started a new life.

It was kind of him to say I did not have to worry about anything, but I knew I could not manage without working. I was interested in the hotel business and he offered me the job of manager. It was a good life and I was happy there. It didn't take me long to master the business. The Ein Hayam was a three star hotel with 75 rooms catering for foreign tour parties, Catholic groups visiting the Holy Land, and Israelis. It was a well-known hotel and always full.

Through my conversations with Max, and several phone calls to Aaron in America, I managed to piece together what had happened to them after the War. Aaron's first wife had perished in the War with their two children. He met his second wife in a concentration camp and they married soon after returning to Lodz, where I had met him. In the summer of 1945, he and his wife left Poland illegally and made their way to a refugee centre in Munich. It was there he met Max, who he had known in Sulejow. He had told Max that I had survived the War with the partisans, but after the War had been killed in Poland. Because of this, when Max returned to Poland to try and trace members of the family, he didn't even bother to look for me. He then left Poland for Vienna, where he made a new home with his family.

Aaron had managed to trace one of his aunts to New York, and went there to set up home. In 1972, he and his wife visited relatives in Israel for the first time. Two days before their return, they decided to take a short break by the sea in Netanya. They had wanted to book into the Hotel Metropole, but there were no vacancies. The receptionist advised them to try the Hotel Ein Hayam, just across the road. There were no rooms there either. But the receptionist told him if they wanted to wait for an hour, a couple was going to the airport and they could take their room once it had been cleaned.

The next day, as they were leaving the dining room, Aaron spotted Max in the reception area. They had a warm reunion and Aaron asked, 'Are you staying in the same hotel as me?" Max replied, 'No. You are staying in my hotel.'

They spent several hours together at Max's home, but my name never came up. It was the next day, while Aaron was visiting Fiszel in Bat Yam, that he found out I was still alive.

Falling in Love

In September 1972, while I was managing the Ein Hayam Hotel, I came out of my office to the reception desk. I saw two guests waiting for the lift. I thought that the woman looked like the Greek singer, Nana Maskouri. The man was older. After they had gone up, I checked the register cards and saw that it wasn't her. Their surnames were different and I assumed they were a father and daughter.

After a short time, I went up to the first floor to the storage room to check a delivery. While I was there, the Greek looking lady came to the door and asked me if I spoke English. I said no, but she managed to explain that she was having problems with the air conditioning in her room and the bathroom door was not closing. I went with her to sort the problem out, and she thanked me. It was Kol Nidre, the day before Yom Kippur.

On Yom Kippur morning, I set up the self-service breakfast table. 'Nana Maskouri' came down to reception and I invited her to help herself to the buffet. She told me she fasted on Yom Kippur. She said she was going to the beach while her father went to the synagogue. I asked her to wait a minute while I went to fetch a plastic bag with a lemon and knife for her, in case she felt unwell in the heat. She thanked me and left.

The following day we got talking, and I found out that she was a widow. Getting to know each other, we fell in love. In February 1973, I visited her and her family in England and asked her to marry me and to come and live in Israel. She said she couldn't leave London because of her elderly father and younger son (who was at college). Her elder son was married, and she had family and friends who she didn't want to be parted from.

I visited London several times during the following year and moved there in 1974. By making my life in London, I reasoned, I would not be so far from my children in Poland, and it was relatively close to Max in Vienna. I went to college in London to learn English and in 1975, Sylvia and I married. Her two sons, Stephen and Ian Annett, accepted me like a father. Stephen was married and working as a journalist and Ian was at university studying Hispanic Studies.

I set up my own business as a 'home improvement' builder and decorator. Five years after our marriage, I took up British citizenship.

In 1978, we tragically lost Ian, who died in an accident in Turkey while working as an English teacher. This affected us all badly and my health started to deteriorate. For Stephen and his wife Kate's, four children – Ben, Leah, Saul and Isabel – I was always grandpa. I am very proud of them.

I always missed my own children, and friends who helped me during the War. Over the years, I followed the news closely and was excited when Solidarity came to power in 1989. The political climate did not enable me to return to Poland until June 1990, when I visited with Sylvia and my grandson Ben.

I visited Kamila, her husband, Krzysztof, and son, Bartosz, in Lodz, where Zbyszek came to see me with his two sons, Marcin and Filip. I wanted to show Sylvia and Ben where I had lived in the forest. We travelled by car to Blogie from Lodz and walked into the forest. I found the well; only a hole was left. My bunker was still visible as a ditch, overgrown with moss and leaves. In Blogie I went to the cemetery to visit Marian's parents' graves. Before I left Poland, I had erected a stone tombstone to replace the birchwood crosses I had made during the War. Janina had asked me to write on the base 'Erected by Florian Mayevski.' She was buried alongside her parents.

I corresponded with Janina after I left Poland. We often wondered what had become of Marian. He left Poland soon after the end of the War, disillusioned with the communists coming to power. Janina had no idea where he had gone.

I have made several visits back to Poland and see people that I knew in the War; I remind them how much I owe them for helping me. When I left Poland in 1969, I felt let down by the country I had fought for and loved. I never generalize about any people. I fought together with Poles and Germans, so I knew that not everyone was a fascist and anti-Semite. I am pleased that, at last, Poland is a truly democratic country. I always believed in multi-party democracy. Like millions of others, I am proud of what I did in the War and feel that I did not fight in vain.

Index